PROBLEMS OF LIFE

AN EVALUATION OF MODERN BIOLOGICAL THOUGHT

by

Ludwig von Bertalanffy

University of Ottawa, Canada
Late Professor of the University of Vienna

JOHN WILEY & SONS INC.
NEW YORK

WATTS & CO.
5 & 6 JOHNSON'S COURT, FLEET STREET, LONDON, E.C.4

COPYRIGHT 1952 BY C. A. WATTS & CO., LTD.

Title of the German Original:
DAS BIOLOGISCHE WELTBILD
Die Stellung des Lebens in Natur und Wissenschaft
A. Francke A. G. Bern 1949

Printed in Great Britain by Richard Clay and Co., Ltd.,
Bungay, Suffolk

CONTENTS

THIS BOOK IS DEDICATED TO

MY WIFE

IT WOULD NOT HAVE BEEN WRITTEN WITHOUT
HER HELP AND DEVOTION THROUGH MANY YEARS

FOREWORD

THE right to speak of a biological world-view follows from the central position occupied by biology in the hierarchy of the sciences. Biology is based on physics and chemistry, the laws of which are an indispensable groundwork for the investigation and explanation of the phenomena of life. It embraces an abundance of particular problems—such, for example, as those of organic form, purposiveness, and phylogenetic evolution—that are alien to physics and make the biologist's research and concepts different from those of the physicist. Finally, biology provides the basis of psychology and sociology; for the investigation of mental activity is based upon its physiological foundations, and similarly the theory of human relations cannot neglect their biological bases and laws. Because of this central position among the sciences, biology has possibly the greatest multiplicity of problems: The phenomenon of " life " is a meeting-place for those conceptions which, according to the usual distinction, originate in the exact sciences on the one hand and in the social sciences on the other.

But the purport of biology for modern intellectual life is even more deeply rooted. The world-concept of the nineteenth century was a physical one. Physical theory, as it was then understood—a play of atoms controlled by the laws of mechanics—seemed to indicate the ultimate reality underlying the worlds of matter, life, and mind, and it provided the ideational model also for the non-physical realms, the living organism, mind, and human society. Today, however, all sciences are beset by problems which are indicated by notions such as " wholeness," " organization," or " *gestalt* "—concepts that have their root in the biological field.

In this sense biology has an essential contribution to make to the modern world-concept. True, it took things rather easily in the past, when it adopted its basic conceptions from other sciences. It borrowed the mechanistic view from physics, vitalism from psychology, and selection from sociology. But its mission, both as a science conceiving and mastering the phenomena peculiar to its own field and as contributing to our basic conception of the world, will be accomplished only by its autonomous development. This is the significance of the striving after new conceptions that has taken place in biology in the past decades.

For more than twenty years the author has been advocating a biological standpoint known as the *organismic conception*. This has been applied to many biological problems in his own work and that of his pupils and co-workers, and of many other scientists who have joined the movement. It has also had considerable influence on neighbouring sciences. Thus, in the theory of " open systems " it unveils new perceptives in physics and physical chemistry ; it leads to new conceptions in the various biological fields and makes the challenge to develop exact and specific laws of organic systems, which laws have actually been formulated in several fields ; it has been used in applied biology, and even in fields so different as, for example, medicine and forestry ; and, finally, it leads to basic philosophical conceptions.

So this book is based on the author's own work, theoretical and practical ; at the same time it gives a survey of results that, scattered among various investigations and publications, have been difficult to grasp as a whole.

We shall see that biology is an autonomous science in the sense that its problems require the development of specific conceptions and laws ; further, that biological knowledge and conceptions are active in different fields. In the present volume we give a survey of basic biological problems and laws within the framework of the organismic conception. From there we proceed to questions of biological knowledge, and eventually arrive at the general principles of the modern conception of the world and the claim for a " General System Theory."

In a volume which is in preparation, and to which some references are made in the text, we shall first discuss in detail some problems that have been outlined in this book. A fundamental problem of biology is that of organic form ; it will be shown, by a survey of the author's research work in the field of " dynamic morphology," that it is amenable to exact investigation and laws. This problem leads to the more general one of " the organism as a physical system," i.e., the theory and laws of the state characteristic of living systems, which is also a new chapter in physics and physical chemistry. On the basis of the biological knowledge gained, we are then able to establish connections with other neighbouring fields— medicine, psychology, and philosophical anthropology. This leads to the questions of the place of man in nature, of symbolism as the fundamental characteristic of the evolution of the human mind, of the relations between evolution and civilization, biology and history, the natural and social sciences. At the same time we shall be enlarging our knowledge of the parallelisms between various sciences and fields of phenomena, and broadening the basis for a detailed presentation of General System Theory as a comprehensive science. Biological, medical, psychological, and anthropological points of view, and that of System Theory, finally lead us to the psychophysical and the problem of reality, with an attempt to overcome the Cartesian dualism of " body " and " mind."

The treatment given here is based entirely on the results of concrete research. The selection of material is determined, however, by the general line of thought, and so is the literary documentation. In the author's *Theoretische Biologie* the interested reader can find a survey of modern biological research.

The author owes the completion of this volume to lovely days in Switzerland. He wishes to express his grateful appreciation to the publishers, Director and Dr. Lang, and recall the friendship that he enjoyed as a guest of President Dr. A. Jöhr, in an environment of Swiss art and culture.

CHAPTER ONE

BASIC CONCEPTIONS ON THE PROBLEM OF LIFE

The youth, attracted by nature and art, trusts, in his vivid desire, soon to enter into the innermost sanctuary. The man realizes, after a long peregrination, that he went no further than into the Propylæa.—GOETHE, Introduction to *Die Propyläen.*

Thus, the task is, not so much to see what no one has seen yet ; but to think what nobody has thought yet, about that what everybody sees.—SCHOPENHAUER.

1. *The Classical Alternative*

AT a time of tremendous upheaval comparable with what we have today, science was presented with an idea that was to influence profoundly man's conception of the world. The time was the Thirty Years' War, and the man who expressed the idea was the French philosopher René Descartes. Impressed by the successes achieved in the young science of physics, then in the throes of its first progress and foreshadowing the possibilities since realized in modern technology, Descartes formulated his theory of the *bête machine.* Not only did the inanimate world obey the laws of physics—this was how Descartes's thoughts went—but so also did all living organisms. Descartes therefore interpreted animals as machines, of a very complicated kind, to be sure, but comparable in principle nevertheless with man-made machines, the actions of which are governed by the laws of physics. True, Descartes was not wholly consistent. A faithful son of the Church, he set a limit to the knowledge of physics : Man was not to be regarded as a mere machine, but as endowed with free will not submitted to the law of nature. Even this limitation was to be overcome by French Enlightenment. In 1748 the Chevalier Julien de la Mettrie set up the *homme machine* against the *bête machine* of Descartes.

These thinkers sought the answer to one of the age-old problems of philosophy. A living organism, plant or animal, is apparently very different from non-living things, such as crystals, molecules, or planetary systems. Life is expressed in an endless variety of plant and animal forms. These exhibit a unique organization proceeding from the single cell to tissues, organs, and multicellular organisms composed of myriads of cells. The life-processes are equally unique. Every living thing maintains itself in a continuous exchange of composing materials and energies. It can respond to external influences, the so-called stimuli, with activities, and especially with movements. Indeed, it frequently shows movements and other activities without any stimulus from outside, and in this we have an obvious, though by no means decisive, contrast between non-living and living things, in that the former are set in motion only by external forces, while the latter can show " spontaneous " movements. Organisms go through progressive transformations, which we call growth, development, senescence, and death. They are produced only by their kin, by the process known as reproduction. In general, the offspring resemble the parents, a phenomenon we call heredity. A survey of the organic world shows, however, that it represents a stream of forms surging up through geological time. These forms appear to be related by reproduction and evolution, changes having occurred in the course of ages, leading to the efflorescence of higher forms from the lower forms. Organic structures and functions are admirably fitted for the " purposes " they serve. An astounding multiplicity of processes goes on even in the simplest cell, so arranged that its identity is maintained in this ceaseless and tremendously complicated play. Equally, every living being displays in its organs and functions a purposeful construction, adapted to the environment in which it normally exists.

If the peculiar nature of living organisms is thus evident—and we hardly ever find ourselves in doubt whether we have a living thing or an inanimate one in front of us—then the question must arise whether or not an intrinsic distinction really exists between the realm of the living and that of the non-living. We ourselves are living beings, so the answer to this question must determine, in large measure, the place we assign to man.

The application of the laws and methods of the physical sciences to the phenomena of life has led to an uninterrupted series of triumphs, both in theoretical knowledge and in the practical control of nature. Descartes initiated the school of physicians and physiologists known to the history of science as the iatromechanics, who tried to explain the function of muscles and bones, the movement of blood, and similar phenomena on the basis of mechanical principles. Harvey's discovery of the circulation of the blood (1628) marked the beginning of modern physiology. Later on the application of acoustics and optics, of the theory of electricity, heat theory, energetics, and other physical fields provided an inexhaustible source of knowledge and helped to explain an ever-increasing number of biological phenomena. Biophysics was augmented by biochemistry. Once it was believed that organic compounds, which are characteristic of living beings, and are, in nature, found exclusively in them, could be produced only in the life-processes. In the year 1828, however, Wöhler produced urea in the laboratory, the first organic compound to be synthesized. Since then organic chemistry and biochemistry have become most important fields in modern science. They also form the basis of chemical industry, from the chemistry of dyes to the hydrogenation of coal, the manufacture of artificial rubber, and the therapeutic armamentarium of modern medicine, including vitamins, hormones, and the chemotherapy of today. At the turn of the century, roughly speaking, came the youngest of these sciences, that connecting link known as physical chemistry, including, for example, reaction kinetics, the theory of the colloidal state and of electrical phenomena in physicochemical processes. It was essential for understanding many life-processes, such as the actions of enzymes, vitamins, hormones, and drugs, the functioning of nerves and muscles, etc.

With the publication of Darwin's *Origin of Species* in 1859, the theory of evolution triumphed. Whereas the great systematist Linnæus had considered animals and plant species as the work of individual acts of the Creator, now an enormous array of facts in all biological fields has been collected to demonstrate that the organic world has climbed, through long generations and geological times, from lower and simpler forms to higher and more complicated ones. At the same time, with his theory of natural selection, Darwin propounded an explanation for this evolution. Every now and then small accidental variations appear in a species. They may be disadvantageous, indifferent, or useful. If disadvantageous, they are soon eliminated by natural selection in the struggle for existence ; if, however, they happen to be useful, they give their owners an advantage in the competition of life, so that they are more likely to survive and reproduce their kind ; thus, in the course of generations, useful variations are preserved and enhanced. Repeated through long ages, this process has led to the evolution of the different forms of living organisms and their progressive adaptation to their environments. Whereas Descartes had pointed to a divine Creator as the engineer of the living machines, now the origin of purposiveness in the living world seemed explained on the basis of chance variations and selection, eliminating all purposive agents.

Thus, the programme put forward by Descartes was the starting-point for developments that not only form the basis of biological science, but also exert a profound influence on human life. Yet in spite of these triumphs, the suspicion has never quite died down that perhaps the very essentials of life have remained untouched and unexplained. Just one year after La Mettrie's *Homme machine*, a polemic pamphlet with the title *Man Not a Machine* was printed in London. The story goes that the author was none other than La Mettrie himself. Should this be true, the chevalier has given evidence of a sovereign irony and freedom of mind that is almost unique in the history of science. In the course of time, this antagonistic viewpoint has been expressed in many different ways. The form most important even now, because it is logically the most consistent, is that given by Hans Driesch (since 1893). Driesch was one of the founders of developmental mechanics, that branch of biology which has as its subject the experimental investigation of embryonic development, and a classic experiment led him to reject the physico-chemical theory of life.

In the greenish depths of the sea, the sea-urchins lead a contemplative life, aloof from the problems of world and science. Yet these same peaceful creatures became the cause of a long-drawn and violent controversy about the essence of life. When a sea-urchin egg begins to develop, it divides first into two, then into four, eight, sixteen, and finally many cells, and in a series of characteristic stages it eventually forms a larva looking somewhat like a spiked helmet and known to science as a " pluteus " ; from this the sea-urchin finally develops by way of a complicated metamorphosis. Driesch divided a sea-urchin germ, just at the beginning of its development, into two halves. One would expect that from such a half-germ only half an animal would develop. In fact, however, the experimenter watches a ghostly performance like that in Goethe's *Sorcerer's Apprentice : " Wehe, wehe, beide Teile stehn in Eile schon als Knechte völlig fertig in die Höhe "*—out of each half comes not a half but a whole sea-urchin larva, a bit smaller, it is true, but normal and complete. The production of whole organisms from divided germs is possible with many other animals. Even identical twins, which occasionally appear in man, are produced in a similar way ; they are, so to speak, a Driesch experiment performed by nature herself. The reverse experiment and other arrangements are also possible. Under certain conditions two united germs produce a unitary giant larva ; by pressing an embryo between glass plates the arrangement of the cells can be severely altered, and still normal larvæ are produced.

B

Like the Sorcerer's Apprentice, Driesch found something uncanny in his experiment, and he came to the
conclusion that physical laws of nature are transgressed
here. Supposing that only physical and chemical forces
operate in the germ, the arrangement of processes which
eventually leads to the formation of an organism can be
explained, according to Driesch, only by assuming that the
processes are directed in the right way by means of a fixed
structure, a " machine " in the widest sense of the word.
But there cannot be such a machine in the germ ; for a
machine cannot achieve the same performance, in this
case the production of a normal organism, when it is
divided, when its parts are dislocated, or when two
complete machines are fused. Thus, Driesch states,
here the physico-chemical explanation of life reaches its
limit, and only one interpretation is possible. In the
embryo, and similarly in other vital phenomena, a factor
is active which is fundamentally different from all physico-
chemical forces, and which directs events in anticipation
of the goal. This factor, which " carries the goal within
itself," namely, the production of a typical organism in
normal as well as in experimentally disturbed development, was called entelechy by Driesch, using an Aristotelian notion. Looking around for purposefully acting
factors, we find their like in our own intentional action.
It is factors that are ultimately comparable with the
mental factors in our purposive actions which make the
crucial difference between the living and the non-living,
and which cause the more than mechanical and physical
properties of life.

In this way we find two fundamental and antithetical
biological conceptions, which in their beginnings go back
to the dawn of Greek philosophy. They are customarily
termed *mechanism* and *vitalism*.

The expression " mechanistic theory " has been used
in widely different senses, a fact that has much encumbered and confused the issue. We have already mentioned
the two most important meanings of this term. First, the
mechanistic conception sees in living things only a

complicated play of those forces and laws which are also present in inanimate nature. A second meaning is seen in the machine-theory of life; the arrangement of events characteristic of all processes in the cell and the organism is interpreted in terms of structural conditions.

In contrast, vitalism denies the possibility of a complete physico-chemical explanation of life and maintains an intrinsic difference between the living and the non-living. It starts, as we have seen in Driesch's doctrine, from the phenomena of regulation, i.e., restitution after disturbances, which seems inexplicable on the basis of a " machine." Other vitalists arrive at their conception by carrying through consistently the machine theory of life. Every machine implies an engineer to design and build it. In this sense, Descartes drew a logical conclusion when he inferred a divine spirit as creator of living machines. Darwin's theory put chance in the place of the creative spirit. Modern biology has made it highly probable that this explanation holds good at least for the origin of varieties and species, and perhaps also for some of the higher systematic units. It is, however, much harder to decide whether it is also sufficient to explain the origin of the great plans of organization and the origin of that interaction of innumerable physiological processes necessary to the functioning of every organism. Locomotives and watches do not usually arise in nature by a play of chance forces—do then the infinitely more complicated organic " machines " ? Thus, the arrangement of the superlatively numerous physico-chemical processes, by means of which the organism is maintained and restored even after serious disturbances, and, further, the origin of the complicated " machine " of the organism cannot be explained, according to vitalistic doctrine, save by the action of specific vital factors, whether we call them Entelechy, Unconscious, or World Soul, which interfere, purposely and directively, with physico-chemical events.

At once, however, we see that vitalism must be rejected as far as scientific theory is concerned. According to it, structure and function in the organism are governed, as it were, by a host of goblins, who invent and design the organism, control its processes, and patch the machine up after injury. This gives us no deeper insight; but we merely shift what at present seems inexplicable to a yet more mysterious principle and assemble it into an X that is inaccessible to research. Vitalism says nothing else than that the essential problems of life lie outside the sphere of natural science. If that were so, then scientific research would become pointless; for even with the most complicated experiments and apparatus, it can lead to no other explanation than the anthropomorphism of primitive mankind, who see an elfin intelligence and will similar to their own in the apparent directiveness and purposiveness in living nature. Whether we consider the behaviour of an animal, or the multiplicity of physical and chemical processes in a cell, or the development of organic structures and functions, we always get the same answer—it is just a soul-like something standing behind them and directing them. The history of biology is the refutation of vitalism, for it shows that always it was just those phenomena which appeared inexplicable at the time that seemed the domain of vitalistic factors. Thus the production of organic compounds was considered a vitalistic phenomenon up to the time of Wöhler; so was the fermentative activity of the cell even for Pasteur, and until Buchner carried through fermentation with yeast extracts at the end of the nineteenth century; so was the phenomenon of organic regulation in Driesch's doctrine. But progress in research has brought an ever-increasing number of phenomena, previously regarded as vitalistic, into the realm of scientific explanation and law. We shall see that even the bankruptcy of scientific explanation, as declared by Driesch in view of embryonic regulation, is by no means unavoidable. On the contrary, the vitalistic argument can be neatly refuted.

The contest between the mechanistic and vitalistic conceptions is like a game of chess played over nearly two thousand years. It is essentially the same arguments that always come back, though in manifold disguises, modifications, and forms. In the last resort, they are an expression of two opposing tendencies in the human mind. On the one hand, there is the tendency to subordinate life to scientific explanation and law ; on the other hand, there is the experience of our own mind, taken as a standard for living nature, and inserted into the supposed or actual gaps in our scientific knowledge.

2. *The Organismic Conception*

In our time a fundamental change of scientific conceptions has occurred. The revolutions in modern physics are widely known. They have led, in the relativity and quantum theories, to a radical reform and expansion of physical doctrine, outranking the progress made in centuries of the past. Less obvious, but perhaps not less significant in their consequences, are the changes that have taken place in biological thought, changes that have led both to a new attitude towards the basic problems of living nature and to new questions and solutions.

We might take as an established fact of the modern development in biology that it does not consent completely to either of the classical views, but transcends both in a new and third one. This attitude has been called the *organismic conception* by the author, who has worked it out for more than twenty years. Similar conceptions have been found necessary, and have been evolved, in the most diverse fields of biology, as well as in the neighbouring sciences of medicine, psychology, sociology, etc. If we retain the term " organismic conception " we shall consider it merely as a convenient denomination for an attitude which has already become very general and largely anonymous. This seems to be justified, in so far as the author was probably the first to develop the new standpoint in a scientifically and logically consistent form.

Biological research and thought have hitherto been determined by three leading ideas, which may be called the *analytical and summative*, the *machine-theoretical*, and the *reaction-theoretical conceptions*.

It appeared to be the goal of biological research to resolve the complex entities and processes that confront us in living nature into elementary units—to *analyse* them —in order to explain them by means of the juxtaposition or *summation* of these elementary units and processes. Procedure in classical physics supplied the pattern. Thus chemistry resolves material bodies into elementary components—molecules and atoms ; physics considers a storm that tears down a tree as the sum of movements of air particles, the heat of a body as the sum of the energy of motion of molecules, and so on.   A corresponding procedure was applied in all biological fields, as some examples will easily show.

Thus biochemistry investigates the individual chemical constituents of living bodies and the chemical processes going on within them.   In this way it specifies the chemical compounds found in the cell and the organism, as well as their reactions.

The classical cell theory considered cells as the elementary units of life, comparable to atoms as the elementary units of chemical compounds.   So a multicellular organism appeared morphologically as an aggregate of such building units ; physiologically, it was the ten 'ency to resolve the processes in the whole organism into the processes within the cells.   Virchow's " cellular pathology " and Verworn's " cellular physiology " gave a programmatic statement of this attitude.

The same point of view was applied to the embryonic development of organisms.   Weismann's classical theory (p. 56) assumed that there exists in the egg nucleus a number of *anlagen* or tiny elementary developmental machines for building the individual organs. In the course of the development, these are progressively segregated by means of the cell divisions the germ is passing through, and thus located in different regions. They bestow on those regions their specific characters, and so finally determine the histological and anatomical structure in the fully developed organism.

Of great importance, not only from the theoretical

but also from the clinical viewpoint, is the classical theory of reflexes, centres, and localization. The nervous system was considered to be a sum of apparatuses established for individual functions. For example, in the spinal cord segmental centres for the individual reflexes are present; similarly in the brain centres for the various fields of conscious sense-perceptions, for the voluntary movements of individual muscle-groups, for speech and the other higher mental activities. Accordingly, the behaviour of animals was resolved into a sum or chain of reflexes.

Genetics considered the organism as an aggregate of characters going back to a corresponding aggregate of genes in the germ cells, transmitted and acting independently of each other.

Accordingly, the theory of natural selection resolved living beings into a complex of characters, some useful, others disadvantageous, which characters, or rather their corresponding genes, are transmitted independently, thus through natural selection affording the opportunity for the elimination of unfavourable characters, while allowing favourable ones to survive and accumulate.

The same principle could be shown to operate in every field of biology, and in medicine, psychology, and sociology as well. The examples given will suffice, however, to show that the principle of analysis and summation has been directive in all fields.

Analysis of the individual parts and processes in living things is *necessary*, and is the prerequisite for all deeper understanding. Taken alone, however, analysis is not *sufficient*.

The phenomena of life—metabolism, irritability, reproduction, development, and so on—are found exclusively in natural bodies which are circumscribed in space and time, and show a more or less complicated structure; bodies that we call " organisms." Every organism represents a *system*, by which term we mean a complex of elements in mutual interaction.

From this obvious statement the limitations of the

analytical and summative conceptions must follow. First, it is impossible to resolve the phenomena of life completely into elementary units; for each individual part and each individual event depends not only on conditions within itself, but also to a greater or lesser extent on the conditions within the *whole*, or within superordinate units of which it is a part. Hence the behaviour of an isolated part is, in general, different from its behaviour within the context of the whole. The action of an isolated blastomere in Driesch's experiment is different from what it is in the whole embryo. If cells are explanted from the organism and allowed to grow as a tissue culture in an appropriate nutrient, their behaviour will be different from that within the organism. The reflexes of an isolated part of the spinal cord are not the same as the performances of these parts in the intact nervous system. Many reflexes can be demonstrated clearly only in the isolated spinal cord, whereas in the intact animal the influence of higher centres and the brain alters them decidedly. Thus the characteristics of life are characteristics of a system arising from, and associated with, the organization of materials and processes. Thus they are altered with alterations in the whole, and disappear when it is destroyed.

Secondly, the actual whole shows properties that are absent from its isolated parts. The problem of life is that of *organization*. As long as we single out individual phenomena we do not discover any fundamental difference between the living and the non-living. Certainly organic molecules are more complicated than inorganic ones; but they are not distinguishable from dead compounds by fundamental differences. Even complicated processes, considered a long time as being specifically vital, like those of cell respiration and fermentation, morphogenesis, nerve action, and so on, have been explained to a large extent physico-chemically, and many of them can even be imitated in inanimate models. A fundamentally new problem is presented, however, in the singular and specific arrangement of parts and processes that we meet with in living systems. Even a knowledge of all the chemical compounds that build a cell would not explain the phenomena of life. Already the simplest cell is a superlatively complex organization, the laws of which are at present only dimly seen. A " living substance " has often been spoken of. This concept is due to a fundamental fallacy. There is no " living substance " in the sense that lead, water, or cellulose are substances, where any arbitrarily taken part shows the same properties as the rest. Rather is life bound to individualized and organized systems, the destruction of which puts an end to it.

Similar considerations apply to the processes of life. So long as we consider the individual chemical reactions that take place in a living organism we are unable to indicate any basic difference between them and those that go on in inanimate things or in a decaying corpse. But a fundamental contrast is found when we consider, not single processes, but their totality within an organism or a partial system of it, such as a cell or an organ. Then we find that all parts and processes are so ordered that they guarantee the maintenance, construction, restitution, and reproduction of organic systems. This order basically distinguishes events in a living organism from reactions taking place in non-living systems or in a corpse.

This has been depicted vividly as follows :

" Unstable substances dissociate ; combustible ones burn occasionally ; catalysers accelerate slow processes. There is nothing extraordinary about this. But that catabolism does not destroy the organism which it is continually nibbling, but on the contrary indirectly maintains it, makes it an organic process. That the constant glow in our tissues does not attack their structure ; that therefore every animal and plant resembles a steam engine made of fuel and yet incessantly working ; in this fact is respiration distinguished from ordinary oxidation.

Just so excretion would be an osmotic phenomenon like any other if it were not for the fact that the glands remove what is noxious for the organism and retain what is valuable. We can easily explain the movements of plants and lower animals as reactions to stimuli; and who is willing to avoid a sharp dividing line across the animal kingdom, will lastly interpret also spontaneous movements in the same way; to him they are ' brain reflexes,' very complicated indeed, but not essentially different from simple reflexes that take place in reaction to external stimuli. Now let us imagine that a dead reflex-apparatus is constructed. It must be charged with latent energies; even slight disturbances would be able to release powerful movements; a special apparatus would provide for continual storage of potential energy. In what way would such mechanism differ fundamentally from a living being, the action on it from a stimulus, its movement from organic movement? In the fact that all organic reactions directly or indirectly serve to maintain existent, or to produce demanded forms " (J. Schultz, 1929).

Thus the problem of wholeness and organization sets a limit to the analytical and summative description and explanation. In what way is it accessible to scientific investigation?

Classical physics, the conceptual scheme of which was adopted in biology, was to a large extent summative in character. In mechanics it could consider a body as a sum, in heat theory, a gas as a chaos, of mutually independent molecules. In fact, the word " gas," introduced in the sixteenth century by the physician van Helmont, denoted just " chaos," in unconscious symbolism. In modern physics, however, the principles of wholeness and organization gain a hitherto unexpected significance. Atomic physics everywhere encounters wholes that cannot be resolved into the behaviour of elements considered in isolation. Whether atomic structure or structural formulæ of chemical compounds or space-lattices of crystals are investigated, problems of organization always arise and appear to be the most urgent and fascinating of modern physics. From such viewpoint, the analytical and summative attitude towards the living seems to be a tremendous solecism. A silly dead crystal has a marvellous architectonic, the design of which makes the mathematical physicist's reasoning work at its utmost speed. But living protoplasm, with its astonishing properties, was thought to have been explained when it was called a " colloidal solution." An atom or a crystal are not the result of chance forces but of organizational ones ; yet it was thought possible to explain the organized things *par excellence*, the living organisms, as chance products of mutation and selection.

The task of biology, therefore, is to establish the laws governing order and organization within the living. Moreover, as we shall see presently, these laws are to be investigated at all levels of biological organization—at the physico-chemical level, at the level of the cell and of the multicellular organization, and finally at the level of communities consisting of many individual organisms.

How is biological organization to be interpreted ?

All knowledge starts from sensory experience. The primary tendency, therefore, is to devise visualizable models. When, for example, science came to the conclusion that elementary units called atoms are at the basis of reality, its first conception was that of tiny hard bodies similar to miniature billiard balls. Not until later was it realized that this is not so, and that the final units are entities not to be defined by visual models, but only by mathematical abstractions, concepts like " matter " and " energy," " corpuscle " and " wave " indicating only certain aspects of their behaviour. When mankind watched the spectacle of the regular movements of the stars, first they looked for powerful machineries, the rotation of which keeps the stars going in harmonic motion—those crystal spheres Aristotle dreamed of—until astronomy destroyed this picture, learning that the order of planetary movements is due only to the mutual attraction of the heavenly bodies in the empty space. Thus structure is the first thing the human mind looks for to explain the order of natural processes ; an explanation in terms of organizing forces is much more difficult.

This applies also to the explanation of life. Observing the inconceivable multiplicity of processes going on in the cell or in the organism, in order to maintain its subsistence, only one explanation seemed possible. It is what may be called the *machine theory*, meaning that the order in vital phenomena was to be interpreted in terms of structures, mechanisms in the widest sense of the word. Examples of this conception are Weismann's theory of embryonic development (pp. 56 f.), or the classic reflex and centre theory (pp. 114 f.) ; but the same type of explanation can be found in every field of biology.

Now structural conditions are to a large extent present in the living organism. The physiology of organs—for example, of organs of nutrition, circulation, secretion, of sense organs as receptors for stimuli, of the nervous system and its connections, and so on—is nothing but a description of the technical masterpiece which confronts us in an organism. In the same way we find structures as mediators of order in every cell, from the muscle and nerve fibrils, as apparatus for contraction and the conduction of excitation, to the cell organs of secretion and division, the chromosomes as structural units of heredity, and so forth.

Nevertheless, we cannot consider structures as the primary basis of the vital order, for three reasons.

First, in all realms of living phenomena we find the possibility of regulation following disturbances. Driesch is right that such regulation, for example, in embryonic development, would be impossible on the basis of a " machine," for a fixed structure can respond to certain definite exigencies only, not just to any one whatever.

Secondly, there is a fundamental difference between the structure of a machine and that of an organism.

The former consists always of the same components, the latter is maintained in a state of continuous flux, a perpetual breaking down and replacement of its building materials. Organic structures are themselves the expression of an ordered process, and are only maintained in and by this process. Therefore, the primary order of organic processes must be sought in the processes themselves, not in pre-established structures.

Thirdly, ontogenetically as well as phylogenetically we find a transition from less mechanized and more regulable states to more mechanized and less regulable ones. To illustrate this again by an example from embryonic development: If at an early stage a piece of presumptive epidermis of an amphibian embryo is transplanted to the region of the future brain, it becomes part of the brain. At a later stage, however, the embryonic regions are determined irrevocably to form certain organs. Thus a piece of presumptive brain will become, even after displacement, brain or a derivative, for example, an eye which develops in the cœlomic cavity, and is here, of course, totally misplaced. A similar fixation to only one function, a progressive mechanization as we may call it, is found in the most diverse phenomena of life.

We come therefore to the following conclusion. Primarily, organic processes are determined by the mutual interaction of the conditions present in the total system, by a *dynamic* order as we may call it. This is at the basis of organic regulability. Secondarily, a progressive mechanization takes place, i.e., the originally unitary action segregates into separate actions, governed by fixed structures. The primary nature of dynamic as opposed to a structural or machine-like order, is seen in fields as diverse as those of cell structures, embryonic development, secretion, phagocytosis and resorption, the theory of reflexes and centres, of instinctive behaviour, *gestalt* perception, etc. Organisms *are not* machines, but they can to a certain extent *become* machines, congeal into machines. Never completely, however, for a thoroughly mechanized organism would be incapable of regulation following disturbances, or of reacting to the incessantly changing conditions of the outside world. The fact that organic processes never represent a mere sum of single structurally fixed processes, but to a greater or less extent always have the character of processes determined within a dynamic system, gives them adaptability to changing circumstances and regulability following disturbances.

The comparison of the organism with a machine also leads to the last of the points of view we have mentioned, the one we call the *reaction theory*. The organism was considered as a sort of automaton. Just as a penny-in-the-slot machine, by virtue of an internal mechanism, delivers an article after a coin has been inserted, so the organism responds to the stimulation of a sense organ with a certain reflex action, to the intake of food with the production of certain enzymes, and so forth. Thus, the organism was considered an essentially passive system, set into action only through outside influences, the so-called stimuli. This " stimulus-response scheme " has been of fundamental importance, especially in the theory of animal behaviour.

In fact, however, the organism is, even under constant external conditions and in the absence of external stimuli, not a passive but a basically *active* system. This is obvious in the fundamental phenomenon of life, metabolism, the continuous building up and breaking down of component materials, which is inherent in the organism and not forced upon it by external conditions. This viewpoint becomes especially important in considering the activity of the nervous system, irritability, and behaviour. Modern research has shown that we have to consider autonomous activity, as it is manifest, for example, in the rhythmic-automatic functions, as the primary phenomenon rather than reflexes and reactivity.

We can therefore summarize the leading principles of an organismic conception in the following way : *The conception of the system as a whole* as opposed to the *analytical* and *summative* points of view ; the *dynamic conception* as opposed to the *static* and *machine-theoretical* conceptions ; the consideration of the organism as a *primary activity* as opposed to the conception of its *primary reactivity*.

These principles enable us to overcome the antagonism of the mechanistic and vitalistic conceptions. Both are based on the analytical, summative, and machine-theoretical principles. The mechanistic theory did not approach just the fundamental problems of life—order, organization, wholeness, and self-regulation. These remained unsolved by analytical investigation, and the attempt to explain them by way of the machine theory, i.e., on the basis of pre-existing structures, leads to failure in dealing with basic phenomena and problems. Vitalism starts with these unsolved problems. But it does not overthrow the summative and machine-theoretical conceptions. On the contrary, vitalism views a living organism as a sum of parts and machine-like structures, assuming them to be controlled and supplemented by a soul-like engineer. Thus Driesch, for example, declared the embryo to be a " sum-like aggregate " of cells, converted into a whole by entelechy. Thus, instead of starting with an unbiased view of the organic system, vitalists also start with the preconceived conception of the organic machine. They realize that, in view of the phenomenon of regulation and of the origin of the machine, this conception is not satisfactory. In order to save it, they introduce factors that repair the machines after disturbance or act as their maker. Thus only two possible explanations of organic order and regulation have been recognized : orderliness through fixed machine-like structures, or as the result of some vitalistic factor. Both are inadequate. The mechanistic view breaks down in face of the phenomenon of regulation and of the origin of the " machine "; vitalism renounces scientific explanation.

Opposed to both, stands an organismic conception. For understanding life phenomena it is neither sufficient to know the individual elements and processes nor to interpret their order by means of machine-like structures, even less to invoke an entelechy as the organizing factor. It is not only necessary to carry out analysis in order to know as much as possible about the individual components, but it is equally necessary to know the laws of organization that unite these parts and partial processes and are just the characteristic of vital phenomena. Herein lies the essential and original object of biology. This biological order is specific and surpasses the laws applying in the inanimate world, but we can progressively approach it with continued research. It calls for investigation at all levels: at the level of physico-chemical units, processes, and systems; at the biological level of the cell and the multicellular organism; at the level of supra-individual units of life. At each of these levels we see new properties and new laws. Biological order is, in a wide measure, of a dynamic nature; how this is to be defined we shall see later on.

In this way the autonomy of life, denied in the mechanistic conception, and remaining a metaphysical question mark in vitalism, appears, in the organismic conception, as a problem accessible to science and, in fact, already under investigation.

The term " wholeness " has been much misused in past years. Within the organismic conception it means neither a mysterious entity nor a refuge for our ignorance, but a fact that can and must be dealt with by scientific methods.

The organismic conception is not a compromise, a muddling through or mid-course between the mechanistic and vitalistic views. As we have seen, the analytic, summative, and machine-theoretical conceptions have been the common ground of both the classical views. Organization and wholeness considered as principles of order, immanent to organic systems, and accessible to scientific investigation, involve a basically new attitude. What occurred to the organismic conception was, however, what usually happens to new ideas: first it was attacked and refused, then declared to be old and self-evident. In fact, once it is realized, this conception merely draws

the consequences from the obvious statement that
organisms are organized. To achieve this unbiased
approach it was necessary, however, and in many fields
is still necessary, even today, to combat deeply rooted
habits of thought.

The organismic conception must be examined, first,
in its significance as a *method of research and theory in
biology* ; secondly, in its *epistemological significance*.

Busy with special questions and experiments, the
research worker in the laboratory is looking at " general
considerations " with mistrust and aversion. Concrete
problems cannot, of course, be tackled by methodological
considerations and postulates but only by patient investigation of the object.   But on the other hand, fundamental
attitudes determine what problems the investigator is
able to see ; they decide the framing of his questions, his
experimental procedure, the choice of method, and
finally, the type of explanation and theory that are given
for the phenomena investigated.   In fact, the dependence
on prevailing attitudes of mind is the stronger the less it is
felt.   In this sense there is no doubt that the work
achieved and the triumphs gained, as well as the short-
comings of classical biology, were determined by those
leading principles which we have indicated.   In order to
realize this, it suffices to glance at any field of biology,
and even of medicine and psychology as we shall see later.
In a similar way, the organismic conception is a working
attitude seeking to direct what problems shall be set and
how they shall be solved.   It makes it possible to see and
to tackle basic problems of living phenomena and their
possible explanations, problems that through the previous conceptions were either not seen at all or, if seen,
considered to be mysteries inaccessible to scientific
approach.

The aim is the statement of *exact laws*, which, according to the essential characteristics of living phenomena,
must to a large extent, have the nature of system-
laws.   In this sense the organismic conception is a
prerequisite for the transition of biology from the stage

c

of natural history, i.e., description of forms and processes in the organisms, to an exact science. It seems to be the task which is set to our age, to accomplish in biology that " Copernican revolution " which, in the sciences concerned with inanimate nature, took place with the transition from the Aristotelian world-system to modern physics.

With this in mind, we will examine some basic biological problems and see how the organismic conception works. Thereafter we shall examine its epistemological consequences.

CHAPTER TWO

LEVELS OF ORGANIZATION

ἓν διαφερόμενον ἑαυτῷ.—HERACLITUS.

It is the basic feature of the living unity to separate, to unite, to commute, to specify itself, to appear and to disappear, to consolidate and to flow, to stretch and to contract.—GOETHE, Aphorismen zur Naturwissenschaft.

1. *Physical and Biological Elementary Units*

WE find in nature a tremendous architecture, in which subordinate systems are united at successive levels into ever higher and larger systems. Chemical and colloidal structures are integrated into cell structures and cells, cells of the same kind to tissues, different tissues to organs and systems of organs, these to multicellular organisms, and the last finally to supra-individual units of life.

At the base of this architecture stand the *elementary units of matter*. According to physics, matter consists of ultimate units, electrons and protons being bearers of negative and positive electricity respectively, and neutrons as electrically uncharged particles. They build the *atoms* of chemical elements. Around a nucleus, consisting of neutrons and protons, the sun of a miniature planetary system so to speak, a number of planetary electrons is revolving, each element being defined by a characteristic number of these particles. Disturbances of this system lead to the transmutation of one chemical element into another and to the splitting of the atom, whereby enormous quantities of energy are released, as demonstrated fatally by the atomic bombs.

Atoms join into *molecules* of the various chemical compounds, belonging partly to the inorganic nature, partly to the organic. While living beings in their constituent chemical elements are not different from inanimate things, organic compounds are specific. From this arises the division into " inorganic " and " organic " chemistry. Among the bio-elements, carbon holds a special place : it appears that life is bound to the ability of carbon to form the most diverse, largest, and most complicated molecules, an ability in which it surpasses all other chemical elements. Justification for a distinction between organic chemistry as the chemistry of carbon compounds, and inorganic chemistry lies in the fact that the number of carbon compounds is a multiple of that of all other chemical compounds taken together. Among organic molecules, the so-called *macromoleculars*, including, for example, the proteins, as the most important building materials of protoplasm, and the cellulose of the cell wall of plants, already show specific structural laws that surpass those of inorganic molecules.

Molecules of small molecular weight, inorganic as well as organic, are defined by precise formulæ, which result from the mutual saturation of the valency bonds of the atoms. As is well known, the number of valency bonds of an element is indicated by the number of hydrogen atoms that it is able to bind. For example, hydrogen has a valency of one, H—H (hydrogen molecule) ; oxygen

two, $O{\big\langle}{\,}^{H}_{H}$ (water) ; nitrogen three, H—N—H (ammonia) ;

carbon four, H—C—H (methane).

In this picture of a molecule defined by a precise structural formula, there is no change, at first, when we pass to organic molecules, apart from the fact that the structures, and hence the formulæ, become more complicated as we look at big molecules like those of chlorophyll, hæmoglobin, vitamins, and the like.

New structural principles do appear, however, as soon as we come to the high-molecular organic compounds. If we consider, for example, cellulose, we find as the smallest unit a disaccharid, the so-called cellobiose; at least three hundred of such " elementary units " are joined by ordinary chemical bonds into a " main-valency chain." But then, a larger number of the latter, about forty to sixty, are further joined by means of secondary valencies or van der Waals forces into a bigger structure, a " micella." Proteins, in contrast to polysaccharides, are distinguished by the fact that they are built up not from equal but from different units, namely, different amino-acids. Within the long chains of protein molecules, definite structural laws are again present, the discovery of which is a major problem of modern organic chemistry. One of these principles is that the different amino-acids seem to be arranged in a periodic pattern. For instance, in the fibroin of silk every second link is glycine, every fourth alanine, every sixteenth tyrosine. According to another structural principle, the molecular weights of many proteins are multiples of a unit of 35,000 molecular weight.

We may state here three points of general significance. First, besides the valencies of classical chemistry, there appears a broader field of forces conditioning the cohesion of matter. They already appear in the so-called imperfect gases as van der Waals forces, caused by the mutual attraction of the gas molecules, and producing the deviations from the ideal gas equations. They play, as lattice forces, a constituting part in the formation of many crystals, and represent, also, the forces of cohesion in solid bodies. As just mentioned, they are operating in the structure of macromolecular organic compounds, and, equally, as we shall see presently, in the formation of pat-terns into which micellæ unite. All these forces are identical, in principle, with those of chemical affinity. The point is, however, that the primary or main valencies of classical chemistry represent only a rather small section of this field. In this sense, structural chemistry, the theory of the gaseous, liquid, and solid states, colloid chemistry, crystallography, etc., fuse into a united realm. At the same time the gulf existing between dead and living structures is lessened. We are gaining an insight into a realm of forces that produce pattern and organization in a field that surpasses the pattern of atoms and atomic groups in the molecule, as considered solely in classical chemistry.

But, secondly, the kind of order is essentially changed. In macromolecular carbohydrates especially, the concepts of the " molecule " and of the " chemical compound " in the classical sense become inapplicable. This is already shown by the fact that, e.g., in the formula for cellulose $(C_6H_{10}O_5)_n$ an undefined number n appears. The structure of cellulose cannot be stated in a rigid formula, but only in a statistical way ; *about* three hundred sugar residues form a main-valency chain, *about* forty main-valency chains a micella.

Thirdly, micellæ again can be arranged in higher structures. Thus, for example, micellæ of cellulose show regular arrangements in the cell wall of plants which in a continuous hierarchy of levels lead finally up to microscopic and macroscopic vegetable fibres. The hierarchy of levels is particularly impressive in the proteins. Amino-acids and protein molecules form parts of higher units, presenting themselves as microscopic fibrillæ ; fibrillæ can again be united to microscopic fibres, and these, in their turn, to macroscopic ones as shown, for example, in nerves and muscles.

Thus the field of *submicroscopic morphology* (Frey-Wyssling [1]) forms a transition from the physico-chemical realm to the biological. Classical physics and chemistry knew only two limiting cases : on the one hand, the rigid arrangement of atoms or radicals in molecular structures and in three-dimensional crystal lattices ; on the other hand, the complete lack of order in the irregular motions of molecules in solutions. But actually the series of physical structures does not end with the

[1] A. Frey-Wyssling, *Submikroskopische Morphologie des Protoplasmas und seiner Derivate.* Berlin, 1938. *Submicroscopic Morphology of Protoplasm and its Derivatives.* Translated by J. J. Hermans and M. Hollander. New York, 1948.

molecular structures and the crystals beyond which only the laws of random distribution in solutions and the molar laws of mechanics apply. Rather there appears a series of higher structural patterns that leads, in an uninterrupted succession, up to the macroscopic realm. At every new structural level the degree of freedom increases. Molecular structure is determined by a precise formula. Macromolecular compounds like cellulose can only be defined statistically. In micellar arrangements like the fibrillæ the building units, in contrast to classical crystals, are ordered not in all three dimensions of space, but only in two or one.

The next step beyond the macromolecular compounds leads into an enigmatic border-land between inanimate and living nature. It is the field of those pathogenic agents that are comprised under the name *viruses*. Infantile paralysis, smallpox, measles, influenza, rabies, foot-and-mouth disease, etc., as well as many plant diseases, are caused by viruses. Viruses are so small that most of them can be made visible not with the usual light microscope but only with the electron microscope. So far as simple viruses are concerned, they can be isolated as pure and crystallized proteins of an enormous molecular weight. For instance, the molecular weight of the tobacco-mosaic virus is 40·7 million. On the other hand, they display a property that seems most characteristic of the living : they multiply by division. If, for example, a plant is inoculated with a few hundreds of molecules of crystallized tobacco-mosaic virus, it shows the disease in all its parts, an enormous increase of the virus substance taking place.

Viruses are, with respect to their chemical and physical properties, the best-known representatives of those entities which can be comprised under the notion of *elementary biological units*. These are defined as being the smallest systems capable of reproduction of their like, or co-variant reduplication. Comparable to viruses in many respects are the units of heredity or *genes*, which are arranged in the chromosomes of the nucleus like a string of pearls, and situated in the strongly stainable sections of the chromosomes, the chromomeres. There is, however, the difference that viruses are introduced from outside, parasite-like, whereas the genes are necessary components of the cell. Genetical research on the one hand and the microscopical study of chromosomes on the other lead to the conclusion that the order of magnitude of the genes is that of large protein molecules with a length of some hundred-thousandths of a millimetre. The chromosome as a whole can therefore be considered as an " aperiodic crystal " (Schrödinger). In ordinary crystals the lattice points are occupied by atoms or radicals which recur in a periodic order, as for example, sodium and chlorine atoms alternate in the crystal pattern of common salt. A chromosome, in contrast, is a crystalline pattern of different groups of atoms, namely, the genes.

The group of elementary biological units comprises viruses, genes, and some other systems capable of reproduction, such as the plasmagenes of the cytoplasm much discussed in recent years, and perhaps also the antibodies. They present three basic problems. The first is the question of the forces that hold together, in a specific way, these systems, which are enormous from the viewpoint of physics, since they contain millions of atoms. A giant molecule is under the incessant bombardment of the molecules of the surrounding solution, which are in thermal motion. In spite of this, a gene is a very stable structure. Genes and chromosomes are transmitted through many generations unchanged, namely, as long as no hereditary variation or mutation takes place. The frequency of mutation of a single gene is of a very small order.

The second question is that of the conditions of growth in these units. Their growth is not comparable with ordinary polymerization as it occurs, for example, in the production of synthetic rubber, where molecules are linked longitudinally so that the chain grows in length, and may eventually break transversely. Instead, the thread-like virus molecule must annex suitable building units transversely, so that eventually a longitudinal division takes place, as we see it directly in chromosome division. However, the ordinary lattice forces of crystals are in no way sufficient to account for the amazing specificity with which elementary biological units pick out of the building materials available just the " right ones " and annex them just at the right places. These specific attraction forces can be directly observed. As is well known, every somatic cell of an animal or plant has a double (diploid) set of chromosomes, i.e., a pair of each sort ; in the maturation of sex cells, meiosis takes place by which the chromosomes are distributed in such a way that every sex cell receives only a single (haploid) set of chromosomes, so that in fertilization by the fusion of the ovum and the sperm cell the diploid set is re-established. A characteristic stage in meiosis is the pairing of the chromosomes. Two chromosomes come to lie alongside each other, the members of a pair often twisting around each other, so that an exchange of sections of chromosomes, and thus crossing over, can take place. The pairing always takes place not only between homologous chromosomes, i.e., the two morphologically corresponding members of the diploid set, but also between homologous chromomeres, i.e., the strongly stainable sections that represent the site of the genes. For example, part of a chromosome with the chromomeres contained therein can, so to speak, be shot away by means of X-rays. If such a chromosome conjugates with a normal one, then, since the pairing takes place between homologous chromomeres, and in one chromosome a section is lacking, a loop is formed by the complete chromosome. As to the nature of the specific attraction forces that are exerted by the elementary biological units and their subsections, only hypothetical statements can be made at present. According to Friedrich-Freksa, they are caused by the pattern of electrostatic charges of the nucleic acid chains, according to Jordan by quantum-mechanical resonance.

The third basic problem is the ability of elementary biological units to perform co-variant reduplication. It has been said : " Before the inscrutability of this mystery of nature we must bow in awe " (Frey-Wyssling). But in fact, a theory of co-variant reduplication has already been attempted in several ways. A hypothetical model developed by the author may be briefly indicated.

Dehlinger and Wertz (1942) have applied to elementary biological units the theory of the " steady state in open systems," as developed by von Bertalanffy. According to them " the simplest arrangement that satisfies Bertalanffy's postulates, i.e., that continually performs chemical reactions and yet in spite of this is quasi-stationary, is a so-called unidimensional crystal, an arrangement that consists of a variable number of molecules (atom groups) packed one on top of the other, into which diffusion takes place from outside, and which is capable of division." This conception was elaborated into a more definite model of elementary biological units (von Bertalanffy, 1944). According to the latter, elementary biological units are crystallites, which, on the one hand, attach molecular groups by virtue of specific attraction forces and thus grow, but which, on the other hand, undergo catabolic processes. If such processes are present, repulsive forces must result which may lead eventually to a division of the crystallite, i.e., to its co-variant reduplication.

The basic assumption in this model of the elementary biological units is that they are not stable crystals, but, like all biological systems, are in a state of continuous exchange of material. For the chromosomes, at least, this conception seems unavoidable on general grounds, and it has been verified experimentally. The chromosomes control the physiological processes in the cell, as is shown by the fact that an enucleated cell is unable to perpetuate life. As bearers of hereditary factors, the chromosomes exert directing influences on the cell and the whole organism. The chemical effects which are caused by the genes demonstrate that the latter are metabolizing units. Experimental evidence leads to the same conclusion. According to the modern conception (Caspersson), the nucleoproteins are the most important centre of protein synthesis in the cell. Hevesy's investigations on the utilization of radioactive phosphorus by the nucleoproteins of the cell show that the nucleic acids are in a state of continuous wearing out and renewal. According to Brachet (1945), this conception is new and important: the substance of the chromosomes appears to be in a state of continuous regeneration and to be the site of a metabolism. Thus, our assumption that the chromosomes are " metabolizing crystals " is doubtless correct. With respect to the viruses, we have indicated methods which may permit an experimental verification.

Two consequences of a general nature follow from this conception. First, genes and chromosomes appear to be not static giant molecules or complexes of such molecules ; rather they are dynamic structures, " metabolizing aperiodic crystals." Their persistence is due to a pattern which is not at rest, but is maintained in a steady state.

Secondly, the much-discussed question whether viruses are " living organisms " or " dead auto-catalysers," may be answered as follows in the above conception. Recent research shows that " virus " is a collective name for entities of a widely different nature. It comprises giant protein molecules such as, e.g., the tobacco-mosaic virus ; bundles of molecules, as the virus of the polyhedra disease in insects ; [1] formations with differentiations visible in the electron microscope, like the vaccine virus ; and finally, forms that already approach bacterial structure such as the waste-water organisms and the rickettsias which cause spotted fever. Limiting the consideration to forms like the tobacco-mosaic protein, it would represent a metabolizing crystallite which is capable of attaching molecular groups from its surroundings, and thus to grow and eventually to divide. But it has not the ability of autonomous life, namely, to carry out the primary syntheses of organic molecules. The necessary con-

[1] Cf., however, G. H. Bergold, " The Multiplication of Insect Viruses as Organisms," *Canadian J. of Research E*, *28*, 1950. (Author's note to the English edition.)

dition for this, a system of enzymes, is certainly lacking. Thus the elementary biological units show growth and co-variant reduplication, but the complete complex of synthetic action, namely, of building up organic molecules *de novo*, remains a privilege of the cell as a whole.

2. *Cell and Protoplasm*

So we come to the cell as the fundamental unit of life. The cell, i.e., a unit whose essential components are nucleus and cytoplasm, is the simplest system known that is capable of autonomous life.   It is a quite astonishing fact that all living beings, from a minute unicellular alga up to a thousand-years-old giant tree, and from the amœba to man, are formed by the variation and multifarious combination of this one building unit.   This fact indicates the presence of a basic structural law.

From the physico-chemical standpoint, protoplasm is an extremely complex colloidal system, wherein the phases, dispersed in water, are represented by particles of widely varying sizes, states of aggregation, physical nature, chemical constitution, and physiological activity ; and in which the degree of dispersion and the transition from sol to more solid gel states are extremely variable and capable of modification.

Obviously, however, even a most complicated colloidal system would still not show the paradoxical behaviour of being " alive," namely, not to enter as fast as possible a state of stable equilibrium, as ordinary physico-chemical systems do, but to remain in the steady state (p. 125) of metabolism.   So we are confronted with the problem of the specific organization of the protoplasm.   A widely accepted modern conception is that of a net-like structure of the protoplasm (Frey-Wyssling) ; submicroscopic particles form an unstable fibrous framework, long thread-like molecules, especially of the proteins, being connected by their side chains at nodal points or " junctions " (*Haftpunkte*).   The *haftpunkt* theory certainly states an important principle of protoplasmic organization and explains many of its physico-chemical properties.

However, a static and structural conception of the protoplasm does not seem to offer a complete solution. We must consider the points of view, expressed earlier in a general way (pp. 16 f.), regarding the limits of explanations in terms of structures.

More than a generation ago, Hofmeister (1901) tried to answer the question of the pattern of processes in the cell. In a liver cell, for example, the size of which is about a hundred-thousandth of a pinhead, a host of complicated chemical processes goes on; processes which the chemist can carry out only by the use of extreme measures, if, indeed, he can carry them out at all, but which the cell achieves neatly by means of specific enzymes. The fact that the reactions go on undisturbed, and in an arrangement that maintains life, was explained by Hofmeister on the basis of the alveolar theory of protoplasmic structure, in vogue at that time : the protoplasm is subdivided by a foam-like structure into tiny separate reaction vessels. The idea that an alveolar or foam structure is fundamental and universal in protoplasm is no longer maintained. However, the basic idea is not changed when we think of the processes separated and arranged through other structural conditions, such as fixation of enzymes in certain microscopic structures or at certain points of a molecular framework by means of chemical, adsorptive, electrical, or other affinities.

Though such fixation of enzymes in protoplasmic structures of a microscopic and submicroscopic magnitude is demonstrated in many cases, it gives no final explanation of the pattern of processes in the cell. The protoplasm is often in an extremely fluid state, as demonstrated by its flow and movement. Its colloidal structure is highly variable. Its volume can be increased by hydration in " cap plasmolysis " tenfold or more ; yet it remains living and capable of reversible dehydration (Höfler). This is hardly compatible with the existence of a permanent molecular framework. Experiments with centrifuged and dissected eggs demonstrate that even a far-reaching separation of the phases and destruction of microscopic and submicroscopic structures of the cytoplasm does not necessarily lead to a disturbance of development. Above all, protoplasm maintains itself in a continuous building-up and breaking-down of its materials, which presupposes an incessant and regulated change of structures.

The organization of the protoplasm is not static but dynamic. The primary orderliness of processes cannot be attributed to pre-established structural conditions. Rather the process as a whole carries its order in itself, representing a self-regulating steady state. Therefore, the system appears to be widely tolerant of disturbances, so long as fixed structural conditions are scarce; but if such conditions are established, as, for example, in embryonic development by the differentiation of the protoplasm into organ-forming regions (p. 58), then their irreversible destruction or translocation leads to irreparable consequences.

That this conception is correct is shown by certain facts, though these are scarce as yet. Probably the individual constituents are present in the cell not as definite chemical individuals but rather in dynamic equilibria with a great number of components. Thus, Sörensen states that the cell proteins have no precisely definable molecules but that they represent a " reversibly dissociable system of components " which, depending on the conditions present, can break into fragments and be re-established from them. Vlès and Gex investigated the transparent sea-urchin egg with the ultra-violet spectrometer. They obtained a spectrum which does not correspond to the protein spectrum. The typical protein spectrum appeared only after destruction of the cell by cytolysis in dilute solutions or by crushing. This result can hardly be interpreted otherwise than that the proteins, considered as stable chemical compounds, are products of a stabilization and are not present in the living cell as such, but rather as components in some sort of dynamic equilibrium. It is also probable that cell structures are, in part at least, not spontaneous, i.e., established on the basis of stable physico-chemical equilibria, but rather that they are non-spontaneous structures needing a supply of energy for their maintenance. A long time ago, the disproportion between the energy visibly consumed in segmentation and that gained through respiration of the embryo was emphasized by Meyerhof. He came to the conclusion that we probably know only partially the work actually done, and that work performed in a realm of small dimensions is necessary to maintain the structure of the cell.

Thus, protoplasmic organization is a meeting-place where the problem of order based on structures flows into that of the maintenance in a steady state. Future theory will have to take into account both aspects.

3. *The Cell Theory and its Limitations*

The statement that the " cell theory " is inadequate is one of the most popular in " holistic " biology. It is necessary to define its meaning clearly in order to give a correct verdict.

The cell, i.e., a system consisting of cytoplasm and nucleus as its essential parts, is the most important structural element in all organisms, plants as well as animals. This is an empirical fact that can neither be contested nor labelled as " theory." This statement, together with all special empirical facts concerning structure and function of cells, can be designated as " cell doctrine."

The " cell theory," however, goes further than this empirical statement. *Morphologically*, it means that the cell is the omnipresent and the sole building element of the living world, and that multicellular organisms are aggregates of cells. *Embryologically*, the development of the multicellular organism was resolved into the actions of the individual cells in the embryo. *Physiologically*, the cell was considered to be the elementary unit of function. Schwann, the founder of the cell theory, had in 1839 already posed the question of whether the totality of the organism determines the growth and development of the cells which are its units ; or whether, on the contrary, the organism is determined by the basic forces of the cells. He decided on the second alternative.

Considering first the morphological statement, the customary assertion that " all organisms consist of cells " is incorrect if expressed in this apodictic way. Complicated protozoa, such as the ciliates, have long been termed " non-cellular." A paramecium, for instance, shows in its single cell organelles analogous to the organs present in higher organisms as multicellular systems : a cell mouth and anus, contractile and neuro-fibrillar structures, locomotive organs, and so on. Thus the cell of a unicellular organism is to be homo-logized only with the multicellular organism as a whole, not with its individual cells. In fact, nature has several times made the experiment of creating larger organisms without cellular differentiation. The *Siphoneæ*, a group of green algæ, are an example. Marine species are often some metres long, have a creeping " stem," finely branched " roots," variously pinnate or crenate " leaves," yet the whole thing consists of a single gigantic multi-nucleate cell. This non-cellular design, it is true, as the rarity of such forms shows, has obviously not stood the test. Cellular differentiation affords an important working structure, giving especially the advantage of a large development of surfaces. Since it is at the surfaces that exchange of material takes place, it is understandable that non-cellular organization is at a disadvantage compared with cellular. Cellular differentiation also facilitates functional differentiation ; while cell membranes and, in another way, the turgor of plant cells, perform an important mechanical function. Considerations of this sort make it understandable that nature holds stubbornly to the principle of cellular construction in the progress towards higher organization. Yet even the higher animal organism does not consist exclusively of cells. Everywhere other structures are also present, such as proto-plasm not organized into typical cells but forming multinuclear masses (plasmodia and syncytia), fibres of muscles and nerves, of the connective tissue, the ground substance or intercellular substance, the body fluids, and so on. Thus a higher organism cannot be simply called a " colony of cells."

Similar considerations apply to the two other aspects of cell theory. The development of the multicellular organism is not a sum of cell actions, but the action of the embryo as a whole, both at the unicellular as well as the multicellular stages. This is shown by regulation, determination, and morphogenetic movements (pp. 43 f., 57 f., 63). Physiologically, the whole of the organism determines the actions of the cells, not the converse. Functions are differentiated not according to cells but according to organs, which can be parts of cells, cells, or complexes of cells (Heidenhain).

4. General Principles of Organization

The architecture envisaged in an organism is typical of a pattern which is of wide occurrence not only in the biological but also in the psychological and sociological fields. It can be called *hierarchical order*. The principles of hierarchical order were defined by Woodger [1] with the aid of mathematical logic.

Hierarchical order in the abstract sense, exemplified by a square divided into four smaller squares, with each of these divided into four still smaller squares, and so on, means : A thing W stands in a relation R to terms or " members " M, to which again subsequent terms stand in relation R. In the example given, R signifies the relation : " to be a quarter of the next superordinate member." A " level " is a class of members that stand to W in the same " power " of R. The following biological examples are indicated by Woodger :

I. The *division hierarchy*, i.e., the four-dimensional order of cells which arises from the division of a cell and

[1] J. H. Woodger, "The 'Concept of Organism' and the Relation between Embryology and Genetics," 1–3. *Quart. Rev. Biol.*, 5/6, 1930–31.—*The Axiomatic Method in Biology*. Cambridge, 1937.

D

its descendants. The relation R (d) here signifies : " to be a direct cell-descendant." W is represented by the mother-cell ; the cells of the first, second . . . etc., generations represent the first, second . . . etc., level of the hierarchy.

There are two classes of division hierarchies : (a) those in which all members, i.e., the cells, are independent organisms (Protista) ; (b) those in which only the first member, namely, the zygote, represents an independent organism, while all other members remain connected, thus forming parts of an organic whole (multicellular plants and animals).

II. In I (b) arises the *spatial hierarchy* of the multi-cellular organism, consisting of an hierarchical order of parts that are connected to systems of ascending order. W here is represented by the whole organism, M are its components, R (s) the organizing relations in which one component stands to a component of the next level.

Among the so-called " parts " of organic systems, two kinds are to be distinguished. A " component " is an assemblage of parts, to which it stands in the relation R (s). Thus nuclei, cells, tissues, and organs are com-ponents, and we can distinguish : (a) components of the cell, (b) cells, (c) cellular components. A " constituent," on the other hand, is a part that lies outside the spatial hierarchy, i.e., that is not analysable into further com-ponents, such as, for example, the ground substance of cartilage or bone, fibrillæ of the connective tissue, blood plasma, yolk, secretory granules, etc. Constituents are always " dead."

The latter definition, however, seems too restricted. What lies outside the division hierarchy of the cells need not also lie outside the spatial hierarchy of the organism. Thus, for example, the intercellular substance of the connective tissue is not a member in the division hier-archy, i.e., it lies outside the relation R (d), but part of it is analysable into hierarchically ordered components, such as systems of fibres of different order, fibrillæ, micellar arrangements, and so on. Not only cell components and cells, but also cellular components may have the ability to divide, as Heidenhain's concept of histo-systems shows (pp. 41 f.). Secondly, the intercellular substance and other formations do not stand outside the organizing relations R (s). Besides the cells, the intercellular substance is also a necessary component in the construction of the higher level of the tissue. In plants especially, we find many " dead " structures, such as cell membranes, cork, tracheids and tracheal tubes, etc., which are nevertheless necessary components of the " living " organism. Chemical and inorganic parts, such as water, hormones, ions, even when they are not components of cells but are located in the body fluids, belong necessarily to the system of the organism, i.e., they partake in the relation R (s). The multicellular organism is not exclusively a cellular hierarchy, though the cells are the ultimate units of autonomous life.

These considerations are important for judging a much-discussed problem of histology, namely, the significance of the intercellular substance. The supporting tissue of the animal organism—connective tissue, cartilage, bone, dentine, and so on—consists in a great part of intercellular substance in which the cells are embedded. We are presented here with two opposing views. According to the first, these substances must be regarded as dead secretions of the cells, which alone are living ; according to the second, the intercellular substance is formed by a transformation of living protoplasm, and the concept of a " living mass " is maintained which includes not only the cells but also the intercellular substance. From the standpoint of the organismic conception, von Bertalanffy (1932) pointed out, first, that growth and morphogenesis of intercellular substances do not suffice to assert their autonomous " life " ; secondly, that their formation is not a sum of individual cell actions, but a unitary action of the whole, often symplasmatic, tissue ; and thirdly, that the conception of a living mass should be replaced by thinking in terms of systems. Within the hierarchical order of the organism, first the cells, then the tissues, are " living," and within the architecture of tissues, intercellular substances play a part similar to that of cell membranes or fibrillæ within the cell, which likewise are not in themselves living, but belong to the system of the cell, which is living as a whole. This organismic view is confirmed by recent developments in histology, especially Huzella's theory of intercellular organization (1941). Classical cell theory, based one-sidedly on the investigation of the structure and function of the individual cells, could not explain how a unitary whole of definite structure and harmonious co-operation arises from the cell aggregate formed by the segmentation of the ovum. On the other hand, a " totalitarian " conception arose as a reaction against the first. It neglected the individuality of the cells, considering cells, intercellular substances, fibrillæ, etc., as a syncytial protoplasmic continuity of the " living mass." The theory of intercellular organization emphasizes, in accordance with the cell theory, that the cells are the ultimate autonomous units of life, and it refuses the notions of a living mass and of extracellular protoplasm. On the other hand, intercellular structures are an important groundwork of the integration and wholeness of the organism. Lifeless cell-products in themselves, in their uninterrupted continuity they act as mediators in the relations between the living cells. According to Huzella, this " elastomotor system " of fibrillæ and membranes, represented in early ontogenetic and phylogenetic stages by the argyrophil fibres (i.e., the system of finest fibrillæ impregnable by silver nitrate), plays, besides its function as a supporting system, a hitherto unknown constructive and integrating role. It forms the medium of life for the cells that it shelters ; it is a depot of nutritive materials and fluids, and a mediator of intercellular relations. It provides a framework for the arrangement of cells and thus for morphogenesis. The extracellular origin, the lifeless nature, and the morphogenetic role of the argyrophil system have been shown in model experiments. Fibrillar and membrane structures can be produced from extracts of connective tissue and can be populated with cells, as well inside the body as in tissue cultures. By allowing salt solutions mixed with the solution of fibrillar substance to crystallize, a fibrillar skeleton of the crystals can be produced. When this fibrillogram is washed to eliminate the salt, and is used as a framework for tissue growth in culture, it is populated by living cells, which take up the arrangement of the original crystal. Finally, the intercellular system is a basis of physiological integration. It gives, for example, an explanation of functional adaptation, since mechanical tension leads to the formation of fibrillæ in definite directions. In the healing of wounds, the liquid filling the cavity of the wound is acidified by inflammation and contains fibrillar substance in solution; the latter gives rise to a fibrillar framework which serves as a roadway for the migration of cells repopulating the granulation tissue. Thus, cellular pathology must be revised from the point of view of the theory of intercellular organization. Disease cannot be reduced completely to disturbances of the individual cells, but is due, to a large extent, to disturbances of the intercellular system. The latter play an important part, for example, in the infiltrating growth of malignant tumours; fibrillæ formed within the connective tissue, which abundantly surrounds the cancer, facilitate and promote the invasion by malignant cells.

III. A third case of hierarchical order which has been analysed by Woodger is *genetic hierarchy*. Here a fertilized ovum represents the first level, with succeeding generations of offspring representing the following levels. The relation $R(g)$ means: "to be a direct descendant." In bisexual reproduction, the genetic hierarchy is, however, only a section of a more complicated order system, which has the character of a network, since a fertilized ovum stands in the relation $R(g)$ to two parents.

IV. Similar principles of organization have been formulated by Heidenhain.[1] According to him, the organism is built of *histo-systems* which are arranged in circles of an

[1] M. Heidenhain, *Formen und Kräfte in der lebenden Natur*. Berlin, 1923.

ascending order, superordinate systems including subordinate ones. For example, in the nerve the following histo-systems are "encapsulated" one into another; neurofibrils, neurons, and, finally, the macroscopic nerve. Histo-systems are distinguished by their ability to reproduce by division. Multiplication by division applies not only to cell components like chromosomes, nuclei, chloroplastids, and, of course, to cells, but also to cellular systems of tissues. When such histo-systems do not separate after division but remain connected, they give rise to systems of an increasingly higher order. Glandular units or adenomeres are an example. Dividing but not completely separating, they show a progressive branching which leads eventually to the formation of a glandular tree. This principle of " division and synthesis " is found in many organs of a glandular type, such as villi, taste-buds, kidneys, etc. It applies also to the leaves of plants, the various forms of which can, according to Heidenhain, be derived by geometrical constructions.

V. An organism displays not only a morphological *hierarchy of parts* but also a physiological *hierarchy of processes*. More accurately stated : an organism does not represent one hierarchy that can be described thoroughly in morphological terms. Rather it is a system of hierarchies that are interwoven and overlapping in many ways, and that may or may not correspond to the levels of the morphological hierarchy. For instance, in the locomotive actions of an animal the following levels can possibly be recognized : first, physico-chemical reactions going on in the muscle ; second, muscular contraction as such ; third, simple reflexes running over definite centres in the spinal cord ; fourth, compound reflexes of larger groups of muscles such as chain-reflexes, activities of synergic and antagonistic muscles, etc. ; fifth, tropotactic reactions, i.e., reflexes that concern the locomotive organs of one side of the body and hence turn the organism into a definite position towards the source of a stimulus ; sixth, reactions of the body as a whole, directed and unified by the highest centres of the nervous system, which co-ordinate all single activities and may also connect them with former experiences ; seventh, social reactions depending on supraindividual units, as, for example, the activities of the individuals in an insect colony.

The hierarchy of processes is much less rigid than morphological organization. It may correspond to the latter if a certain process concerns a morphologically defined component, but not necessarily so. Certain components, for instance, the bulk of the tissue of the pancreas gland and the islets of Langerhans, together constitute a higher component, in this case the organ known as the " pancreas." But with respect to other relations, one component may co-operate with another which is far removed morphologically, to form with it a functional system of a higher order. For example, the islet cells co-operate with the liver to regulate, by means of insulin, the liberation of sugar into the blood.

This is of considerable importance, for it follows that there are " organs " that do not constitute morphological units. Whereas classical anatomy was based on morphological architecture, modern anatomy prefers presentation in terms of " functional systems " (Benninghoff). Systems of action, such as the locomotor system consisting of bones, muscles, and nerves, are intelligible only in the way they interact. It can even be said that just the most important advances in modern anatomy consisted in the discovery of such functional systems, for instance, of the reticulo-endothelial system and of the pacemaker system of heart-beat by Aschoff and others.

VI. A further important type of hierarchical order can be called *hierarchical segregation*. The most obvious example is seen in embryonic development. The developing egg, which originally represents a unitary system, progressively segregates into individual " fields," determining, first, complexes of organs, subsequently, individual organs, parts of organs, and so forth. Thus ectoderm and endoderm are formed within the embryo as a whole. In the former, the regions of presumptive epidermis and medullary plate are established ; within the presumptive medullary plate are established the brain and spinal-cord regions ; within the brain area, the primordium or *anlage* of the eye, and so on. In terms of hierarchical order, the thing W corresponds here to the originally unitary egg, the following levels to the segregates of the first, second, etc., orders. It is important to note that segregation does not coincide with the cellular organization within the division hierarchy. It arises, in mosaic eggs (p. 58), within the still undivided ovum, in regulation eggs, within an already multicellular complex. The factors that determine the future fate of the different regions do not arise because the developing egg, by segregation, is subdivided into cellular components ; rather they are the dynamic antecedent that determines a group of cells to become a certain component. This is shown, for example, in regulation experiments, where removal, displacement, addition of cells, etc., do not alter the segregation of the primordia of the organs.

Segregation hierarchy is especially characteristic of the biological, and also the psychological and sociological realms. Hierarchical order in physical systems, as represented, for example, in a crystal by its space lattice, is formed by integration to a higher unit of initially separate systems, in this case atoms. By contrast, in the biological sphere a primary whole segregates into sub-systems. This is found in embryonic development. Phylogenetically, also, the progressive differentiation of organisms means the segregation of life-functions, which were originally combined in a single cell, into separate systems for the intake of food, for digestion, response to stimuli, reproduction, etc. Similarly in the psychological field. Classical association psychology (pp. 189 f.) assumed that individual sensations, corresponding to the excitation of individual receptor elements, for example of the retina, are experienced primarily, and that they integrated secondarily into perceived shapes. Modern research, however, makes it probable that an unsegregated and, as it were, amorphous whole is primary, and that this differentiates progressively. This is shown, for instance, in pathological cases. With patients recovering after injuries of the cerebral centres, it is not single sensations that re-appear first: a punctiform light causes the sensation not of a luminous point, but of a vaguely circumscribed brightness; only later is the perception of shapes, and finally of points, progressively restored. Comparable to embryonic development, the restoration of vision progresses from an undifferentiated to a differentiated state, and the same probably holds for the phylogenetic evolution of perception.

VII. The general concept of hierarchical order needs to be completed in various directions.

First, the *closeness of interaction* in organic systems is of differing strength. In primitive metazoans, for example in coelenterates, cells show widely independent migrations and phagocytic activity. In contrast, in higher animals we find a strict subordination of cells and tissues to the whole. We may call this progressive integration. The higher we go in the scale of organisms, the more different is the behaviour of isolated parts from that which they display in the whole; and the poorer it is in comparison with the performance displayed by the whole organism.

In higher animals are three main integrative systems. First, the body fluids, which distribute nutrients and oxygen among the tissues and organs, and, at the same time, provide a *milieu interieur* optimal for the functioning of cells. Secondly, the hormones, which regulate the functions chemically in a specific way. Thirdly, the nervous system, which is an apparatus not only for response to environmental stimuli but also for the integration of the organism.

The progressive integration goes hand in hand with the progressive *differentiation* of the parts, which at the same time means specialization, metaphorically spoken of as " division of labour." The fundamental activities of metabolism, growth, irritability, reproduction, heredity, and so on are found in the simplest unicellular organism as well as in the highest animals. Whereas, however, in an amœba all these processes are carried through by one and the same system, namely, the proto-plasm of its cell, in a higher organism they are distributed among different organs and systems. Specialization alone makes the enhancement and refinement of functions possible ; but, on the other hand, it must be paid for.

Progressive differentiation means, at the same time, progressive *mechanization*, i.e., the splitting of the originally unitary action into a sum of separate individual actions, and thus a loss of regulability. When certain parts take over one function more or less exclusively, the ability to regulate, i.e., to take over other functions in the event of emergency, disappears. Hence loss of parts leads to irreparable damage. This principle can be illustrated best by a sociological analogy. In a primitive community of savages, every one is farmer, craftsman, soldier, hunter at the same time. Progress in cultural accomplishments is possible only by specialization of the members of the group in a craft. But then, the specialist becomes irreplaceable, and he is also much more helpless outside his usual occupation than the primitive individual. Thus, Robinson Crusoe cuts a much poorer figure in the wilderness than Man Friday, and he only just manages to exist because a good fairy washed all sorts of civilized things ashore for him. The same is true in the biological field, as shown in the progressive determination of the embryonic regions in development, or in the nervous system, where regulability diminishes with progressive differentiation and fixation of centres. Progress is pos-sible only by differentiation and specialization within the organism itself, as well as with respect to adaptation to different environments. It must be paid for by mechaniz-ation, the fixation of the parts to a single function, and hence the loss of resilience in the face of disturbances.

Moreover, with increasing differentiation certain parts gain predominance over others. It is therefore linked with increasing *centralization*. Thus we find in a more highly developed hierarchy a principle of rank-order and subordination of the parts (A. Müller). In the cell it is the nucleus, in higher animals the nervous system, that is the "central" organ, i.e., that upon which integration particularly depends. Of course, an organism does not, like an army, represent an unequivocal rank-order but rather a complex of diverse and interacting patterns of order. For instance, the brain may be considered as the central leading organ. Yet the brain immediately becomes incapable of acting if the heart stops for a few seconds. The heart, again, cannot be considered the organ of the greatest vital importance, for it fails when the liver does not deliver the sugar necessary for its functioning, and the liver in its turn is again dependent on the correct functioning of the heart (von Neergard).

The principle of rank-order and "leading parts" is also of a general character, and applies not only to morphological organization but to many other fields as well. Thus, for instance, embryonic development is controlled by certain areas, the organizers. Organic catalysers also show a rank-order (Mittasch): beginning with enzymes, which are adjusted most specifically to carry through a single reaction, to biocatalysers, such as the growth-substances in plants or the organizer substances in animals which regulate more or less wide complexes of processes, and up to directing biocatalysers, such as many hormones, that influence to a large extent the whole organism psycho-physically. Similarly, there is a rank-order of the genes, from genes that control single, often minute, characteristics to those that influence a larger number of characters in more or less extensive pleiotropism (p. 74), and finally to "superordinate" or "collective" genes (E. Fischer, Pfaundler) that direct the activity of numerous other genes. To the latter belong the genes of sex determination ; those which control the hereditary variation of the vertebral column, and cause correlated changes not only of the bony system but also of the musculature, the nerve-supply, and so on (Kühne); and possibly genes that govern the types of constitution in man.

Connected with the principle of centralization are the problems of biological individuality.

5. *What is an Individual?*

Observing, under the microscope, the animalcules in a drop of pond-water, we may well ponder about the question: What is an "individual"? This question, apparently somewhat superfluous, is, in truth, profound and difficult to answer. In the drop of water we see a confusing bustle of tiny transparent creatures. Green spindle-like things propel themselves through the water with a long flagella, slipper-shaped creatures move more pompously by the beat of their cilia, amœbas, shapeless droplets of protoplasm, creep around in the mud.

Now it is obvious that a fish, a dog, or a human being is an individual. We mean by this that it represents a living being that is in space, in time, and in action distinguished from others, and as such passes through a definite life-cycle. But in unicellular organisms the notion of the individual becomes muddled. Through many generations, they multiply merely by division. Individual means something " indivisible "; how can we call these creatures individuals when they are in fact " dividua " and their multiplication arises precisely from division? The same holds good for asexual reproduction by fission and budding, as is found in many of the lower metazoa. In the face of experimental evidence, the term " individual " becomes inapplicable. Can we insist on calling a hydra or a turbellarian worm an individual, when these animals can be cut into as many pieces as we like, each capable of growing into a complete organism? Other experiments with fresh-water polyps also demonstrate the extreme vagueness of the notion of the " individual." It is easy to produce a double-headed polyp by making an incision at the anterior end. Afterwards the two heads compete: if a water-flea is caught, both heads quarrel about the booty, although it does not matter at all which one takes it—in any case it goes down into the common gut, where it is digested and so benefits all parts. The question whether we have to deal with " one " or " two " individuals here becomes meaningless. Nature answers it, however, when the double animal either divides into two or fuses into one.

But the notion of the individual is problematic even in the higher animals, at least in early stages of development. Complete animals develop not only from halves of a segmenting sea-urchin egg, as in Driesch's experiment, but also from a segmenting newt's egg. On the other hand, an " individual " can even consist of portions put together from different species. Thus, by fusion of two half gastrulæ, Spemann bred a well-developed newt, one side of which was striped newt, the other a cross between striped and crested.

Finally, from the physical standpoint, the individuality even of man can, on occasion, be questionable. Identical twins originate from a single ovum developing in an early stage into two " individuals." It is well known that there is, in identical twins, an amazing similarity not only of their physical, but also of their mental properties. In pairs of criminal twins, the sort of crimes and the time at which they were committed by both brothers were found to agree in a surprising way.

Thus, from the standpoint of natural science, we can speak of individuality only in the sense that phylogenetically and ontogenetically a progressive integration takes place, the parts of the organism becoming increasingly more differentiated and less independent. Strictly speaking, there is no biological individuality, but only a progressive individualization, both phylogenetic and ontogenetic, which is based upon progressive centralization, certain parts gaining a leading role and thus determining the behaviour of the whole. Individuality is a limit which is approached but not reached, either in development or in evolution.

With individualization death enters the world of the living. Experience shows that systems of the complication and integration reached in higher animals, that are incapable of reproduction by fission as opposed to the primitive " dividua " in the lower phyla, are incapable of unlimited existence ; by natural wear and tear they decline into old age and death. Not inappropriately the individual could be defined in terms of death. There is a direct antagonism between the centralizing tendency of the systems of integration, especially the central nervous system, and the disintegrative tendency of the reproductive organs (A. Müller). Complete individuality, i.e., centralization, would make impossible reproduction, which presupposes the construction of a new organism out of parts of the old. On the other hand, it is precisely the main central systems—brain and heart—which break down first in the natural course of senescence, and are therefore the organs of death.

Thus the notion of the individual is, biologically, only to be defined as a limiting concept. Indeed, it originates in a sphere quite different from that of science and objective observation. Only in the consciousness of ourselves as beings different from others are we immediately aware of individuality that we cannot define rigidly in the living organisms around us.

6. *The World of Supra-individual Organizations*

Organisms confront us as entities that are distinct in space. They are, however, members of higher units of life. In respect of time there is the unit of the species. Every organism is begotten by its like, and may itself give rise to new organisms ; thus, each is a member of a supra-individual unit. In space, also, the hierarchy of life does not end with the organism, and there are still higher units.

To these belong first associations of organisms of the same species, such as colonies of animals. A well-known example is seen in the siphonophores, pelagic colonial hydrozoa, forming mighty stocks that consist of numerous polyps differentiated into feeding, tentacular, float-like, and reproductive units. This case is almost unique in the animal kingdom, but supra-individual organizations are possible also with organisms that are separated in space, as in the societies of insects, ants, bees, and termites. The animals of different castes, workers, sexual forms, soldiers, appear to be subordinate " organs " of the insect society in much the same way as the different sorts of polyps are in the coherent siphonophore colony. The function of the specialized animals is co-ordinated for the maintenance of the whole, no less than is that of the cells and organs within an organism. The whole determines, in the nuptial flight, in swarming, in the bringing up of new queens, the action of the individuals in the hive, an admirable " purposiveness " which far surpasses any possible foresight of the individuals concerned. The whole is preserved and regenerated when, for example, after the death of the queen, new queens are raised in the hive. Thus all criteria of wholeness apply to the insect society. The phylogenetic trend leading to the most highly organized animal societies is comparable to that leading to higher organisms ; here also we find first loose assemblies, with a progress to increasingly higher and more differentiated organizations.

Systems of a higher order can result from the union not only of organisms of the same, but also of different species. Then we speak of symbiosis. Here also there is a series of stages, from a loose living-together, such as that of hermit-crabs and sea-anemones, to those extraordinarily intimate relations where lower organisms are reared often in specially adapted organs of a higher animal. Biology has demonstrated the wide distribution and the importance of intra-cellular symbiosis in its different forms, such as alimentary and respiratory symbiosis, symbiosis with luminescent bacteria, etc. In some cases a new organism originates from the symbiosis of two different ones, as in lichens, which are a symbiosis of algæ and fungi.

From living units that result from the association of organisms of the same, or the symbiosis of different species, we come to still higher systems. The animal and plant communities (biocœnoses) in a certain area, such as a lake or a forest, are not mere aggregates of many organisms, but units ruled by definite laws. A biocœnosis is defined as a " population system, maintaining itself in dynamic equilibrium " (Reswoy).

The highest unit is the whole of life on earth. If one group of organisms were eliminated, it would have to attain a new state of equilibrium or collapse. Green plants alone can synthesize organic matter from inorganic compounds by using the energy of the sun's radiation. Only the co-operation of the various groups of micro-organisms guarantees the maintenance of the cycles of the bio-elements. Similarly, very specific compounds such as the vitamins, which are indispensable for the normal functioning of the animal, are produced only by plants. The stream of life is maintained only in the continuous flow of matter through all groups of organisms.

The theory of biocœnosis belongs to those fields where the notion of wholeness has been applied for a considerable time, as shown by the work of Friederichs, Woltereck, Thienemann, and others. Here only a few general points shall be mentioned.

Biological communities are systems of interacting components, and thus display the characteristic properties of systems, such as mutual interdependence, self-regulation, adaptation to disturbances, approach to states of equilibrium, etc. However, their degree of integration is, of course, small in comparison with organisms; they are loose, uncentralized units. Their development is determined by external conditions, whereas the life-cycle of an organism is determined by internal conditions. Hence biocœnoses are rightly denominated as systems, but they cannot be called super-organisms, as has often been done.

Being biological systems, biocœnoses are governed by exact laws. In fact, the mathematical theory of population systems, of the struggle for life, and of biological equilibria (Lotka, Volterra, d'Ancona, and others) belongs to the most advanced fields of mathematical biology, although experimentation has not yet fully kept pace with theory in this field. Laws can be stated for the growth of populations consisting of a single species, as well as for populations consisting of several species, co-existing in the struggle for life by competition for food or in a predatory-prey relation.

The conception of wholeness as applied to biological communities has not merely theoretical, but also a highly practical significance. Virgin nature is in a state of biocœnotic equilibrium. Though there is a struggle for existence in every seemingly peaceful scrap of forest or pond, there is a balance between plants and animals, prey and predators. No species can increase without limit, because each one has its natural enemies, but none is extinguished as long as there are no genetic or environmental changes. This situation is altered if man carelessly interferes with the biological equilibrium. He cultivates the land and creates associations that consist only of one species of plants, for example, monotonous forests of pines; he introduces involuntarily from other parts of the earth, pests whose natural enemies are missing. This can lead to far-reaching disturbances of the biological equilibrium. Not controlled by their natural enemies, the pests increase unrestrictedly, reaching plague proportions that utterly destroy plantations over wide areas. Biological pest control, which tends to restore the biological equilibrium by the introduction of the pest's natural enemies, has shown with surprising results the practical consequences of the theory of biological equilibrium.

In this connection it is interesting to note that the organismic conception was applied to a very unexpected field, namely, forestry. Lemmel (1939) derived from von Bertalanffy's organismic conception the *dauerwald* (permanent-forest) idea, i.e., the principle of avoiding forest clearances and of preserving the natural biocœnosis as far as possible.[1]

The quantitative laws of population growth also have practical importance. The growth of a population consisting of one species presents, in the case of man, the fundamental problem of social politics. Indeed, the development of mathematical biocœnology was incited by Malthus's doctrine on the growth of human populations.

[1] Recently identical principles have been urged in U.S.A. forestry, in order to restore the catastrophic effects of careless deforestation. (Author's note to the English translation.)

E

The laws of the cycles of epidemics, which can be considered as biological equilibria between man, pathogenic organisms, and carriers, is of considerable importance for hygiene. The question of the factors, environmental or lying within the dynamics of the population systems themselves, that cause periodic fluctuations of the numbers of animals, is of economic importance for hunting, fisheries, agriculture, and forestry.

However, a profound problem arises, for philosophical consideration. Is it justified to consider a biocœnosis as a unitary system? Are its members not engaged in a continuous struggle, annihilating or being annihilated? This leads to ultimate questions of philosophy. A continuous struggle of parts, to use Roux's expression, is active in all biological systems, be they organisms or biocœnotic systems. Not only in a " *dividuum*," such as a budding polyp, do the components compete for building material which each one strives to grasp; the same is true in every living system. Thus, in a starving animal, tissues of less vital importance are consumed to maintain the more important ones; in regeneration or metamorphosis they are recklessly sacrificed in the service of the whole; and even in normal development, the differential growth of the parts, a process basic in morphogenesis, is due to their competition for material. Thus unity by competition of parts is present in every biological system, organisms as well as supra-individual units of life. It is a reflection of that deep metaphysical concept which goes back to Heraclitus and Nicholas of Cusa: The world as a whole, and each of its individual entities, is a unity of opposites, a *coincidentia oppositorum*, which, however, in their opposition and struggle, constitute and maintain a greater whole. From these biological facts, the vista opens to the age-old problem of theodicy, and of the evil in the world that arises from individuation into competing parts, a struggle connoting annihilation of the individual and progressive realization of the whole.

CHAPTER THREE

THE UNITARY CONCEPTION OF THE PROCESSES OF LIFE

There will be a time when the mechanistic and atomistic conception is entirely overthrown in clever brains, and all phenomena will appear as dynamic and chemical, and will thus testify ever more the divine life of nature.—GOETHE, Tagebuch, 1812.

1. *Embryonic Development : the Approach to the Organismic Conception*

FROM a tiny, nearly homogeneous drop of protoplasm rises the marvellous architecture of an animal with its billions of cells, its innumerable organs and functions. In the mysterious process of embryonic development we are confronted with one of the major problems which have been from ancient times at the centre of the theory and philosophy of organic life.

It may be said that an organismic conception has today become the general attitude in this field. But since it has been the main battlefield in the fight between the classical theories of mechanism and vitalism, and since the news of the later skirmishes seem, as yet, not to have got round to the philosophers, it may be just as well to give a short account of the main phases of this controversy.

The present author originally advanced his organismic conception in connection with the problems of embryonic development, which was then the most disputed field in theoretical biology. His book, *Modern Theories of Development* (1928, 1934), tried to establish the necessary course of biological theory by an investigation of the different conceptions, arranged in disjunctions and exhausting the logical possibilities.

From the dawn of modern science there have been two fundamental conceptions to explain the marvellous performance of development. Called " preformation " and " epigenesis," they date back to the beginnings of biology in the seventeenth century. The old preformationists assumed that the human or animal organism is already present in miniature in the ovum or sperm. Like the blossom from the bud, or the butterfly from the chrysalis, it needs only to unfold and enlarge in order to form the developed organism. There are remarkable pictures from the early times of microscopy, showing a tiny mannikin huddled together in the sperm, even with a night-cap on its head. Progress in microscopic technique soon disclosed the incorrectness of this view. Ovum and sperm exhibit few structures, and certainly not those of the fully developed organism. Therefore, the school of the epigeneticists invoked a formative agent or *nisus formativus*, that produces the organism out of the amorphous mass of the fertilized ovum. Of course, this classical form of the theories of preformation and epigenesis was rather naïve. Nevertheless it indicates the two basic schemes of explanation of development which have not lost their significance even now. Preformation contends that, if not the organism as such, at least primordia or *anlagen* for its parts are pre-established in the ovum. Epigenesis considers the ovum as originally undifferentiated.

Thus we find a *first logical disjunction*. Either the visible complexity of the organism is, in an invisible form, already present in the ovum so that it is built up by the action of separate pre-existing *anlagen*—preformation—or this complexity gradually arises in the course of development, this being an action of the embryo as a whole— epigenesis.

The first alternative is presented by the modern preformistic theory advanced by Weismann (1892). According to him, there exist primordia or " determinants " for the different tissues and organs in the nucleus of the fertilized ovum. By means of genetically unequal nuclear divisions, these determinants are progressively distributed among the cells arising in the course of cleavage. Finally, each kind of cell contains only one sort of determinants, which gives to the tissue or organ its character.

Weismann's theory was soon refuted by the first results

of the experimental physiology of development, or developmental mechanics. For instance, half an embryo should contain only one-half of the determinant material, and should produce, therefore, half an organism only. Actually, in many cases, for instance, those of the sea-urchin or the newt, a whole organism develops from each half of a divided embryo. In general, the actions of the cells in regulation, transplantation, regeneration, etc., widely exceed their actions in normal development. Hence embryonic cells and parts are not fixed or preformed for a definite developmental action, as Weismann supposed. Their developmental " potencies " are, in general, far greater than their actual performances in normal development. In an early stage and with certain restrictions, the embryo represents an " equipotential system," i.e., every part can produce all and the same, namely, a complete organism.

This leads to a second question. In what way is it determined which of the potencies of the cells are realized in each particular case? The answer is given in a principle stated by Driesch : The developmental action of a cell depends on its position in the whole developmental system. This statement summarizes an enormous number of experimental results. At present it suffices to mention regulations in early development : Normally, each of the two blastomeres into which a sea-urchin ovum divides, produces the right or left half respectively of the larva. Separated, however, each of these blastomeres produces a complete larva. And if two two-celled germs are fused, each of the four cells contributes one-quarter of the resulting larva.

The question how a certain course in development, or so-called determination, is established is answered by a second fundamental principle of physiology of development. This is the principle of Spemann : Embryonic parts become determined progressively towards a certain developmental fate. This principle also could be illustrated by numerous examples, of which we mention only one here. Prospective epidermis material of a newt embryo, i.e., material which, in normal development, would form ventral epidermis, if transplanted at an early stage into the area of the medullary plate of another embryo, becomes part of the brain, and *vice versa*. At a later stage, however, determination has already occurred. Then a piece of prospective medullary plate, transplanted into the ventral region of the body, produces a portion of nervous system or one of its derivatives, such for instance as an atypical eye.

Therefore development is not based upon a distribution of preformed *anlagen*; instead the determination of embryonic parts towards a definite developmental fate occurs progressively and in dependence on the whole. Development is therefore epigenetic in principle, although preformistic traits are not lacking. No developing system is completely homogeneous, but there are differences, even in the earliest stages, at least along a polar axis, as shown, for instance, by the irreplaceability of the organizer in the newt embryo (p. 66).

These statements give the explanation for a characteristic difference found in the behaviour of animal embryos. In one class, the so-called regulation eggs, examples of which are seen in sea-urchins, newts, and mammals, normal organisms develop from divided or otherwise disturbed eggs or germs. In the other class, the mosaic eggs, for example, of the sea-squirts and molluscs, fragments, such as one-half of the egg, produce only defective organisms. Regulation and mosaic eggs are not contrasting classes; all transitions between these types are found. The difference is based on the fact that, in regulation eggs, the speed of determination is small, while in mosaic eggs it is great, in comparison with the speed of segmentation. Hence, in regulation germs, cells and areas are not determined in the early stages, and so normal organisms may arise even after disturbances. In mosaic eggs, on the other hand, determined regions, the so-called organ-forming areas, are already present before segmentation begins, so that regulation after disturbance is impossible.

So the first alternative is answered. Development is not the action of independent dispositions or developmental machines; it is governed by the whole.

Now a *second disjunction* arises. The " whole " is either a factor different from, and added to, the material system of the embryo, or it is immanent in the constellation of the material system. The first alternative is vitalism, the second a scientific theory of organic wholeness.

We have already seen (pp. 5 f.) how Driesch was led to vitalism by his experiments. In this context, the epistemological and methodological objections against vitalism do not interest us, only its empirical refutation.

Experience shows that the " whole " on which determination depends is not the typical result to be reached in the future, but the actual state of the developing system at a given time which can be indicated in any particular case. To be sure, inasmuch as determination has not yet taken place, there is equifinality (pp. 142 ff.), that is, the same end-result can be reached from different initial conditions. However, development does not proceed " purposively " in the sense that the best and most typical result possible is achieved, as should be the case with an entelechy directing events in foresight of the goal. What really happens, whether, when, and how regulation occurs, is unequivocally determined by the conditions present. For example, $\frac{1}{2}$-blastomeres of the sea-urchin yield complete larvæ, and so do $\frac{1}{4}$-blastomeres; from parts of the eight-celled and later stages complete or defective formations develop, depending on the cell material present or experimentally combined, and it can be indicated in every case which result will be obtained in a given cell combination. It has been said that the course of development proceeds with the " senseless industry of necessity," irrespective of whether the result is good or bad, teleological, dysteleological, or ateleological. Also it cannot be maintained that entelechy would try to achieve the most typical result possible, and is prevented from attaining this goal by the inadequacy of the material available. For example, in super-regeneration up to six hind legs can be developed in toads if suitable incisions are made. Obviously the work of entelechy is not limited here by the lack of material at its disposal; rather the process is necessarily determined by the material conditions present. The force of this argument is enhanced by remembering that, according to Driesch, one of the fundamental actions of entelechy is " suspension " of potential processes, meaning that it stops certain processes in normal as well as in regulative development, in such a way that the most nearly typical whole is formed. Super-regenerates and other monsters clearly demonstrate the impotence of entelechy.

In the second disjunction, therefore, we can exclude the assumption of a principle which joins the material system of the embryo, and governs its action in dependence on the typical result to be reached in the future. The " wholeness " manifest in the processes of development is immanent in, not transcendent to, the embryo. The embryo represents primarily a unitary system, not a sum of developmental machines or *anlagen*, which is the common basis of both Weismann's theory and vitalism, the latter assuming only that the mechanism is guided by a non-spatial entelechy.

Hence comes the *third disjunction*. The unitary action of the embryo can either be explained by application of principles and laws known from inanimate nature or it is specifically organic.

A comprehensive theory of the first kind was developed first by Goldschmidt.[1] According to it, development is essentially based upon chemical actions of a catalytic nature which are inaugurated by the genes and lead to a differentiation of the egg-cytoplasm and later on of the cell regions in the embryo. In turn, the differentiated sorts of protoplasm are localized in a definite pattern, due to physico-chemical equilibria, and thus organ-forming areas are established by the way of " chemo-differentiation." Up to the establishment of primary chemo-differentiation, the embryo represents a unitary physico-

---

[1] R. Goldschmidt, *Physiologische Theorie der Vererbung.* Berlin, 1927; *Physiological Genetics.* New York and London, 1938.

chemical system. Therefore, in regulation eggs, the state of equilibrium is restored after disturbances, and regulation is achieved without making necessary the intervention of an entelechy. In mosaic eggs, on the other hand, chemo-differentiation occurs in very early stages, and thus no regulation is possible after disturbances. With the progress of development, ever new genes react with the chemical differentiations already present, and so a relatively small number of gene-catalysts and organ-forming materials is able to set going a vast number of reactions and morphogenetic processes. The basic principle is that of harmonized reaction-velocities : in every cell and every embryonic area many different reactions are going on simultaneously ; gene actions and their quantitative harmonization decide which of these processes shall take the lead and thus determine the developmental fate of any given part.

Modern research has shown that this picture is essentially correct, though we are still far from being able to define the chemical factors of development and the equilibria to be assumed. Gene-hormones show that the action of the genes is of a chemical nature (p. 76). That fundamental processes in development are controlled by chemical substances of a hormone-like nature is shown by the organizers, as investigated by Spemann and his school. The formation of the nervous system, one of the most fundamental processes in vertebrate development, is induced by a certain region, namely, the dorsal lip of the blastopore, representing the material of the future notochord and mesoderm. Pieces of this organizer from a newt embryo, and of corresponding regions of other vertebrates, transplanted to an atypical place, for instance, to the future ventral epidermis, induce the formation of an atypical, parasite-like nervous system and associated organs. This action is of a chemical nature. Organizers that had been killed, as well as extracts from very different animal tissues, and finally a number of chemical substances can also induce the formation of a neural tube, though the chemical nature of the organizer substances is not yet definitely established. Further, the principle of harmonized reaction-velocities is of far-reaching significance. It was first stated by Goldschmidt for the more special problem of sex determination, but it applies generally to the realization of the genetic pattern (p. 76 f.), and it gives an explanation for many processes in embryonic development: progressive determination, organizer action, self-differentiation contrary to prospective significance (*bedeutungsfremde Selbstdifferenzierung*), polarity, development of bilateral asymmetry, compensatory growth, heteromorphosis, etc., and finally for the harmonization of growth rates both in its ontogenetic and phylogenetic aspects (pp. 99 f., 138).

This, however, does not mean that we may expect a complete resolution of development into physico-chemical factors; rather the opposite is true. The more chemical factors are extricated, the more the problem concentrates in the vital organization of the developmental system. For instance, the insight into the chemical nature of the organizer action shifts the problem to the substratum that is reacting; the organizer appears largely as a trigger agent, the burden of the morphogenetic process falling on the competence of the tissue affected. If we call the egg a " polyphasic colloidal system " we have said very little; a reaction between chemically defined " gene-hormones " and " organ-forming substances " would produce only chemically defined compounds, not organized formations as they are produced in the course of development.

Secondly, we find, in development, highly mysterious processes that appear to be independent of physico-chemical differentiations. It is the production of highly complicated forms from chemically undifferentiated materials. In pure culture, as it were, we find this phenomenon in the development of slime moulds. Here uniform amœboid cells, previously independent of each other, obey, so to speak, a command coming from the unknown : they fuse to form multinucleate protoplasmic streams, and finally build, by organized migrations, complicated and often very beautiful sporangia. In mushrooms, irregularly growing hyphæ seem to fill a mould, as it were, which is invisibly predetermined and characteristic of each species, and so build up the pileus. In sponges and hydroids we find singular phenomena of reintegration; in hydroids fragments produced by chopping the animal up, and in sponges, even separated cells into which the animal is dissociated by pressing through fine silk, unite to rebuild an animal of normal shape and organization. In such processes we find no chemo-differentiation, no separation of organ-forming materials, no chains of chemical reactions which run parallel, and out of which certain are favoured and therefore victorious in the competition, according to the principle of harmonized reaction-velocities. We can understand that differentiation leads to the effect that an originally simple geometrical form—say, a spherical cell aggregate—develops into a more complicated shape, because growth is increased at certain points by growth-promoting factors localized there. But in the mushroom the process is just the reverse : not that an internal differentiation is converted into the production of a more complicated shape, but an originally undifferentiated material showing no internal pattern, assumes progressively a simple shape. If we look at animal development, we find, in general, that material differentiations are intimately connected with morphogenetic movements, as it is the case in gastrulation, the formation of the neural tube, etc. But even here, the same principle is operative. Nearly homogeneous material—nerve-cells, muscle-cells, bone-cells—builds a typical formation, the outer shape of which is characteristic and specific, while the inner arrangement of the cells is largely accidental. Morphogenetic movements appear to be an integrated action of the whole ; with respect to gastrulation, it was spoken of " amœboid movements," not of the individual cells but of the embryo as a whole.

Necessary, therefore, as the isolation of physicochemical processes is, the problem of organization and of the elaboration of shape remain, nevertheless, un-solved.

The postulate of physico-chemical explanation of development may also be made in a more general way. Instead of demanding a factual explanation for special phenomena, we may only postulate that reduction to the laws of physical *gestalten* is possible in principle; *gestalten* meaning systems that attain a state of equilibrium and represent physical wholes (cf. pp. 192 f.). But here too we meet with peculiar difficulties.

Embryonic development from the scantily differentiated ovum to a highly organized multicellular structure connotes an increase of order due to factors lying within the system itself. From the point of view of physics, such behaviour seems at first paradoxical. A physical system cannot increase its order by itself; on the contrary, the second law of thermodynamics demands that in every closed system a decrease of order is the natural course of events. This is exactly what happens in a decomposing corpse, but the process in a developing embryo is just the reverse. The behaviour of the latter presupposes, first, that there are specific organizational forces working towards higher levels of order; and secondly, that the embryo is not a closed system. To make possible increase of order, a continuous supply of energy is necessary, which is used to produce order, but is dissipated in part according to the entropy principle. Following on what was said before, the organization present in the embryo cannot be interpreted in a preformistic and structural way, but only as a dynamic order. From the viewpoint of energetics, development implies the expenditure of work, which is supplied by the oxidation of reserve-materials, such as yolk, present in the egg (cf. p. 127).

A second fundamental characteristic is presented by the historical character of organic systems, which we shall consider later on (p. 111). It is the phylogenetic accumulation of dispositions that unfold progressively in ontogenesis. This historical characteristic also is foreign to inanimate systems.

Thus the last alternative leads to the conclusion that embryonic development presents problems that are beyond a mere application of *gestalt* principles known in inanimate nature. Rather " a specific *gestalt* principle immanent in the organism " is to be supposed, to quote an earlier formulation by the author (1928). This conception is not vitalistic, for it does not assume any transscendent factors interfering in living nature, but on the contrary excludes such factors. But it is organismic, since the organization immanent to living systems is considered to be specific, and so the autonomy of biological systems is contended.

In spite of an enormous amount of experimental data, we do not have, at present, a really satisfactory theory of development. It can be said, however, that all modern conceptions are organismic in the sense just explained : Child's gradient theory, the theory of biological fields, as advanced by Gurwitsch and, in another form, by P. Weiss, Dalcq's field-gradient theory, and so on. Without entering into the details of facts and hypotheses, we shall give an outline of the modern picture of development, which may be considered as well established.

The notion of the " potencies " of an embryonic part can be used in a descriptive way, meaning an enumeration of what this part may become under different conditions. In this sense, for example, ectoderm has the " potencies " to form epidermis and nervous system, and possibly also mesodermal organs. But we must be aware that the notion of potency has no meaning in a realistic sense. In fact, this notion is based upon Aristotelian metaphysics, which was static and dualistic. *Potentiâ*, there are many figures slumbering in the marble, and the sculptor brings one of them into actual existence. In a similar way, organic material seems full of " potencies "; some of the sleeping ones are " awakened," others " stopped." Taking this conception for granted, hardly any other view is possible than that this business is done by an entelechy that is comparable to the genius of the artist. But then it seems paradoxical when such performance is

achieved, in organizer action, by a dull chemical substance, or when, according to Child's gradient theory, a
mere difference in the intensity of metabolism determines
whether a piece of a polyp or of a flatworm shall produce
differentiated organs of the head or the much simpler
ones of the posterior end. A characteristic of this conception is that it considers living things as essentially
inert. The substratum of development appears as a dead
material needing an entelechial artist to fashion it.

In fact, however, the developing embryo is a restless
dynamic flow. The so-called " potencies " of the different
regions and cells are doubtless nothing else than an expression of the fact that, according to the principle of
harmonized reaction-velocities, different reaction trains
are going on in each of them. In the beginning, none
of these reactions has any decided advantage, apart from
an ever-present gradient along the main axis. The latter
is shown, for example, in the newt embryo by the early
determination of the organizer, which is irrevocably determined to form chorda-mesoderm, while all other regions
are capable of producing organs of a widely different
nature, in certain experiments even transgressing the
boundaries of the germ layers. At this stage, and with
the limitation imposed by the existence of axial differences, the system is " equipotential." There is a singular
condition of equifinal pseudo-equilibrium (p. 142 f.), to
which the system returns after disturbances. So it
yields the same typical result, even after division, fusion,
or displacement of parts. Similarly, transplanted parts
develop " neighbourwise," i.e., according to the region
where they were transferred.

Gradually, however, a definite set of reactions gains
the lead in each region. At first, this is still small and
not irrevocable—the state of "labile determination" is
reached : in the normal site, development continues in
the determined direction—prospective epidermis, for
example, becomes epidermis; but this unstable determination can still be altered by additional factors. Then
another set of processes gets the lead, so that the same

piece, transplanted, may yield part of the brain. The same applies when the embryonic part in question is cultivated in a non-specific medium such as a salt solution. Then the material develops " self-wise," i.e., according to its origin : ectoderm to epidermis, mesoderm to notochord, somites, pronephros, and muscle-fibres, endoderm to intestinal epithelium. But slight and seemingly non-specific influences are sufficient to deflect the course of development. Thus, ectoderm, transferred into the orbit of an older newt larva, may yield notochord or musculature by differentiation in a direction contrary to the normal (*bedeutungsfremde Selbstdifferenzierung*) ; if transferred into the body cavity or the lymph spaces, it may yield epidermis or nerve-tissue. Specific influences, originating in certain regions, determine neighbouring areas. This applies especially to the organizer region ; for the formation of a typical nervous system it is necessary that the region of future chorda-mesoderm underlies, during gastrulation, the dorsal ectoderm.

If a certain chain of reactions has gained a definite advantage, then displacement into a new environment is no longer able to stop it. Determination has occurred, and the parts in question are irrevocably fixed to a definite line of development.

In such a way the developmental system, unitary at first, segregates into subordinate systems or " fields." In the course of development, these fields become increasingly autonomous, their boundaries, vague at first, become progressively definite. Thus, the primordia of the organs are formed. In the beginning they are not visibly different, and reveal themselves only in the " self-wise " differentiation of parts in the case of experimental transfer into abnormal environment ; later they become manifest as organ-forming areas. Some of these regions appear to contain all necessary conditions for their future development within themselves ; others, the organizers, exert definite influences on neighbouring areas. The same process goes on within each of these regions. Thus, the primordia of organs, such as those of the heart, the eye, and the extremities, are at first equipotential, only a polar axis being established. Therefore, normal formations can arise from divided, fused, or dislocated primordia; gradually, however, different partial regions are determined. The cells respond to the developmental conditions arising in the different areas by their histological differentiation into diverse sorts of cell; also the course of their histological differentiation goes through the stages of non-determination, unstable, and final determination.

Similar principles apply to regeneration. First we have equipotentiality and non-determination in the young blastema so that we get regulation following division, fusion, and transplantation experiments. This is succeeded by progressive determination, organizer action by means of more or less specific stimuli, and, finally, segregation into autonomous developmental systems occurs. Certain phenomena in this field illustrate especially well the "race" between different reaction-chains in the same material. For example, the stalked eye of a crayfish will regenerate the eye if a cut is made in such a way as not to injure the eye-ganglion; but the stump produces an antenna ("heteromorphic" regeneration) if the ganglion is removed. Obviously here "eye" and "antenna" reactions run parallel, the former getting the lead only with the help of an influence coming from the ganglion. A piece of a polyp forms a head with tentacles when it projects freely in the water; but it forms a foot if buried in the ground. The higher intensity of respiration in the former case gives the "head reaction" the advantage needed. Generally, determination can be interpreted as originally quantitative differences in the velocities of reaction-chains in the various parts becoming qualitative differences when one of these reactions has got the upper hand. Such an advantage may be due either to the gradual increase of an originally small phase-difference in the part itself or to the action of more or less specific factors coming from outside.

Thus, development appears not as the awakening or

inhibition of mysterious "potencies," but rather as a dynamic interplay of processes. Up to now, developmental mechanics has offered, however, only qualitative notions, not a " theory " in the strict sense, a system of statements from which laws and predictions can be derived in a quantitative way. With respect to one partial process of development we are in a more fortunate position : Organic growth is accessible to theoretical conceptions from which exact quantitative consequences can be derived (pp. 136 ff.).

2. The Gene : Particle and Dynamics

Modern genetics is a field of biology which is surpassed by no other in the subtlety of analysis, the integration of originally independent lines of research, the exactness of laws and predictions, and the far-reaching practical applications. In fact, the modern insight into the material substratum of heredity is not unworthy to be compared with the vistas modern physics has opened into the elementary units of matter, the structure and organization of the atoms.

The history of genetics offers many points of epistemological and methodological interest. It is frequently said that progress in biology is possible only " by way of experiments." This statement is, of course, correct in so far as experiment represents the most important tool to unveil the inter-connections of natural phenomena. Mendel's laws, for example, would never have been discovered by arm-chair speculation. However, Mendel's achievement was more than merely carrying out a set of experiments which were very careful indeed, but in no way exceedingly difficult. After all, crossing experiments have been made since man began to cultivate plants and to breed animals, as far back as the upper Palæolithic. There was quite a number of botanists engaged with hybridization experiments before Mendel. What Mendel's achievement actually means is that he started his experiments with a novel and ingenious abstraction. He broke with the conception prevailing up to then,

F

namely, that of "blending inheritance," meaning that in hybrids paternal and maternal properties are in some way mixed, and he introduced, instead, the conception of hereditary characters which are separate and pass unmixed through the generations. This theoretical conception was the prerequisite that exact laws of the distribution of characters in successive generations, following from the doctrine of combinations, could be derived from the experiments. Not haphazard experimentation, but the combination of theoretical ideas with planned experiments has made possible Mendel's achievement and the subsequent advance of genetics. Science is not a mere accumulation of facts ; facts become knowledge only when incorporated into a conceptual system. The pioneering character of Mendel's work lies in the fact that he applied this method, which was ever used in physics, but was so far unheard of in biology ; but this was also the reason that he was not understood even by the best of his contemporaries.

Genetics also gives a refutation of an objection frequently raised against the possibility of exact laws in biology, namely, that the phenomena of life are much too complicated to admit of such laws. The answer is, first, that even physical systems, such as an atom or a crystal, are far from being "simple" ; that physics also makes use of idealized abstractions such as those of the absolutely rigid body or the ideal gas—abstractions, which are corrected successively in the development of theory by further approximations ; and that apparently simple problems, such as the three-body problem of mechanics, can often be solved only by way of approximations. On the other hand, genetics shows with respect to biological phenomena that the complexity of the processes concerned need not exclude exact laws. The train of events that lead from the gene in the fertilized ovum to the characters of the completed organism is far from being known. Nevertheless, we can make perfectly exact statements on heredity by using a suitable abstraction, namely, considering the distribution of characters in the successive generations. This is sufficient to make possible the theoretical and also, in animal and plant breeding, the practical control of heredity.

There is, moreover, one further condition to make this procedure practicable. Research must advance from the simpler to the more complicated phenomena. Poincaré once said : " If Tycho had had the knowledge of modern astronomy, Kepler would not have been able to state his laws " ; an exact knowledge of the disturbances in the planetary orbits would have made it impossible to establish these laws, which are valid only in the first approximation. Similarly, genetics was in the very fortunate position that it proceeded from an especially favourable case—Mendel's pea-hybrids—to increasingly complicated ones. Mendel himself was embarrassed when, in his second investigation, he made crossing experiments with *Hieracium*, which represents, as we know today, a complicated case of interspecific hybrids, and therefore does not obey the simple Law of Segregation. The way of Science is from ideal cases, for which a simple law can be enunciated, to the progressive inclusion of complications. It may well be that in many biological fields we know not too few but too many facts and that the very accumulation of an enormous amount of data hampers the discovery of the necessary theoretical schemes.

Genetics demonstrates in a striking way how irrational factors, beside logic and planning, are active in the advancement of science. Actually, the progress of genetics, and especially the transition from simpler to the more complicated cases, was much favoured by a series of lucky incidents. The first was already present in Mendel's experiments with peas. Mendel analysed seven pairs of factors, and the haploid chromosome number of the pea is just seven. Fortunately the genes of the characters investigated by Mendel happen to be localized each in a different chromosome. If Mendel had experimented with characters due to genes localized in one chromosome and therefore linked, he would not have found segregation, and would have been unable to state the classical laws. A further lucky chance was in the selection of the fruit fly Drosophila as a model-organism to be used in genetical research. Even the choice of *Drosophila melanogaster*, and not of another species, *Drosophila virilis*, for instance, was especially fortunate for carrying through genetical analysis. Again, after an exhaustive genetical analysis of Drosophila had been made, and chromosome maps, based upon a tremendous number of experiments, were established, the significance of the giant chromosomes in the salivary glands was recognized. Giant chromosomes happen to occur exclusively in flies; they made possible the direct cytological evidence for the arrangement of the genes in the chromosomes, as inferred from the genetical experiments. Finally, Drosophila proved to be an especially fortunate object also for experiments in the line of gene-controlled substances, for transplantations, and so on.

Genetics demonstrates that the hereditary characters of organisms are determined by material units that are located in the chromosomes, namely, the genes. They control all hereditary characters, from trifling minutiæ, hardly bearing on the process of living, such as the colour and shapes of pea seeds or the colours of eye and hair in man, to serious defects, such as deaf-mutism or epilepsy, up to highly intellectual characters like musical talent or scientific aptitude. The genes are arranged in the chromosomes in linear series, roughly comparable to a string of pearls. They are freely combinable and are largely independent of each other. According to the third Mendelian law, the chromosomes, with the genes they contain and the corresponding characters, can be freely combined and distributed, according to the laws of probability. Similarly, genes located in the same chromosome have a great measure of independence. In crossing-over an exchange of sections of two homologous chromosomes with their contained genes takes place. In deficiency, individual genes or sections of chromosomes are lost. In translocation and inversion, i.e., attachment of a piece of a chromosome with the contained genes to another chromosome, or inverse insertion of a certain section within a chromosome, the arrangement of genes in the chromosomes is altered without affecting the characters controlled by the respective genes. Apart from rather exceptional cases, namely, the so-called position effect (pp. 81 f.), genes appear therefore as independent units, exerting their influence irrespective of their combination and arrangement. They are arranged serially in the chromosomes, like the wagons of a goods train, which can be shunted and re-arranged without any influence on the load they contain.

This elementalistic conception of heredity is confirmed by an enormous amount of genetical evidence, including both hybridization analysis and direct microscopic investigation of chromosome structure. If, however, we look at it in a quite naïve and unbiased way, it seems in some manner paradoxical.

Take, for example, the pet object of the geneticist, the fruit fly Drosophila. Its chromosomes appear to be stuffed with genes concerned with subtle and, in general, rather insignificant characters. The system of genes or genome seems to consist of units, of which one is concerned, say, with a certain deformation of wings, the next with a certain colour of the eye, another with the formation of some bristles or the colour of the body. No arrangement can be found that corresponds to the organization of the animal, which is certainly not a mere aggregate of eyes of such or such shapes or colours, of wings, and bristles. Thus, it is a fundamental question—though it is usually evaded in the text-books—what the " gene " or " hereditary unit " really signifies.

In fact, an organismic conception is also indispensable in the field of heredity, and the modern developments in genetics are tending that way. Here too, it is necessary to come from static to dynamic conceptions : inheritance is not a mechanism where genes are connected machine-fashion with the visible characters they produce, but rather it is a flow of processes in which the genes intervene in definite ways.

Classical genetics had already accumulated many cases where two or more genes co-operate in producing a character. Recent research has shown that such co-operation is not an exception, but rather the rule and an important characteristic of inheritance. This fact is expressed by the notions of the polygeny of characters and pleiotropism of genes. In the last resort, all inherited characters are polygenic, that is, they are dependent on the co-operation of many or all hereditary factors present. There is a continuous intergradation from characters apparently dependent on the state of one single gene only through characters which are influenced by a smaller or greater number of genes, the so-called modifiers, and finally to characters influenced by many or all genes present, i.e., the gene-complex. On the other hand, gene action is pleiotropic, that is, a single gene affects not only a single character but more or less the whole organism. Here also is a continuous series, from cases where the action of a gene is manifest only in one character, up to genes causing profound changes in the organism as a whole.

It is a consequence of the dynamic nature of inheritance that the same character may be influenced by very different factors. Thus different genes, often situated in different chromosomes, can produce very similar characters. In so-called phenocopy the same effect can appear, on the one hand, as an hereditary mutation due to the alteration of a gene, and on the other hand as a non-inherited modification due to external factors. In Drosophila, for instance, many variations, known as mutations, can also be produced as modifications or *dauermodificationen* by heat treatment. In butterflies treatment with heat or cold in the pupal stage produces modifications that correspond to hereditary subspecies of southern or northern regions. This is not difficult to understand. The number of processes involved in the production of an organism and its characters is extremely large; nevertheless, there is only a finite number of possible alterations, and therefore gene mutations or external factors may lead to phenotypically similar or even identical results.

In order to define the notion of the " gene " we must be clear about what is actually established in genetical experiments. Such experiment is possible only with organisms that can be hybridized. Therefore, ascertained genes always concern characters that are different in the hybridizable organisms, that is, in general racial characters and only in exceptional cases species characters. Genetical analysis shows, for example, that at a certain chromosome locus there exists a certain difference between a Drosophila of the mutation *Bar* and one of the Wild type; a difference which, in this case, can be directly verified by observation of the giant chromosomes under the microscope. Because there is a difference in, and different reactions are induced by, this locus, the one race will develop *Bar*-eyes, having a smaller number of ommatidia, while the other develops normal eyes. Hence what is determined as a " gene " is not a unit or *anlage* producing by itself a definite character or organ, such as this or that colour or shape of eyes, wings, bristles, and the like; rather it is the expression of a difference between genomes that correspond in general. The *whole* organism is produced by the *whole* genome,[1] though in a somewhat different way, depending on the nature of a macromolecule—a so-called gene—at a certain chromosome locus. This is directly shown by the fact that the presence of a complete genome is necessary for normal development. Larger deficiencies, that is, loss of sections of chromosomes, are always lethal. So the genome is not a sum or mosaic of independent and self-acting *anlagen*, but a system that, as a whole, produces the organism, the development of which, however, is altered according to changes in parts of this system, the genes.

[1] And the plasmon, i.e., the totality of hereditary factors in the cytoplasm though the bearing of cytoplasmic transmission appears to be rather limited (mostly cases in higher plants, but also plasmagenes in yeast, bacteria, paramecium, drosophila, etc.). (Author's note to the English translation.)

How do genes act ?   We have an extensive knowledge about gene-dependent substances, that is, hormone-like compounds, the formation of which depends on definite genes and which are active in developmental processes. The action of the genes is of a catalytic nature ; by means of gene-dependent substances they control the rate and direction of processes.   Since a process in development can sometimes be influenced in a similar way either by a mutated gene or by an external factor such as temperature, the same phenotype may appear as a mutation, a phenocopy, or a *dauermodification* ; in a similar way as a chemical reaction may be influenced in the same sense by a catalyzer or by temperature.

The action of many genes is that of " rate-genes," that is, factors that influence the velocity of certain chains of reactions.   Development is based upon a system of gene-controlled processes.   Their correct timing guarantees normality ; on the other hand, mutation of a gene can lead to a change in speed of the reactions it controls, and hence to more or less far-reaching alterations of the organism.   This is the " principle of harmonized reaction-velocities " (Goldschmidt), which we have already encountered in embryonic development.   This principle was first stated by Goldschmidt with respect to sex determination.   The latter is based on the fact that in every organism " male " and " female " reactions go on simultaneously ; the quantitative ratio of the sex factors (in the typical case, two X-chromosomes in the female, one in the male) decides which wins the race.   In genetics, as well as in developmental physiology, numerous phenomena are governed by the same principle.   Thus, dominance can be expressed in terms of reaction velocities ; in the organism a race between concurrent reactions takes place, and the faster one wins.   For this reason dominance can be influenced by other genes (modifiers, the gene-complex) and also, in phenocopies, by external factors, both of which can affect the speed of the developmental processes concerned.   In a similar way, change of dominance can be explained ; a reaction that was slower at the beginning gains the upper hand at a later stage. A shift in the synchronization of early processes of segregation, for example, of segmentation, can produce far-reaching effects, as is the case in the mutation *aristapedia* of Drosophila, where a bristle of the antenna is transformed into a tarsus. Again, if the ratio between developmental rate and rate of chitinization in insects is altered by mutation, larval organs may become chitinized and thus perpetuated in the adult. Conversely, certain organs may develop prematurely, and then, for example, a caterpillar develops antennæ. *Hemmungsmissbildungen*, such as cleft palate, hare lip, and the like, are based on the fact that certain processes of development have proceeded with insufficient speed, so that an embryonic condition is carried into the adult stage. Insufficient hormone-production due to mutation can lead to neoteny, that is, sexual maturity in the larval stage, as in perennibranchiate amphibians like Proteus. A similar explanation applies to " fœtalization " in man according to Bolk's theory (cf. Vol. II). Changes in the rate of chemical reactions can alter pigmentation, and so lead to hereditary colour-varieties. Alterations in the harmonization of the rates of relative growth constitute one of the most important factors in the origin of the multitude of forms in the living world and in evolution (pp. 99 f., 138).

In many cases it can be proved, or at least made probable, that differences in the rate of developmental processes are based upon quantitative differences in the respective genes and are proportional to the gene-doses. This is true, to begin with, in the classical case of Goldschmidt's theory, namely, sex determination. The quantitative ratio of the sex factors, namely, whether one or two X-chromosomes are present, normally determines whether development shall proceed along the male or the female line. In hybrids between different geographical races of the gypsy moth (*Lymantria dispar*), however, the normal quantitative ratio is disturbed. If a " weak " and a " strong " race are crossed, the quantities of the sex factors do not match; the genotypical sex is unable to outrun completely its opposite; and this results in the formation of sexual intermediates (intersexes); the result of every possible combination can be foretold. Similarly, multiple somatic alleles, such as the *vestigial* series of Drosophila (with wings scalloped in a different degree), can be interpreted as quantitative differences in the mutated gene. The mutations *Bar* and *ultra-Bar* of Drosophila (p. 82) consist in a duplication or a triplication of a small section of the X-chromosome visible in the giant chromosomes of the salivary gland. Similarly, quantitative ratios are decisive for the determination of sex in algæ, and lead in certain cases to so-called relative sexuality; in this case the sex-determining substances have been identified chemically.

Many important consequences follow from Goldschmidt's principle. The genetical phenomena already mentioned, as well as the corresponding phenomena of competition in developmental processes (pp. 62, 66 f.), show that differences that are at first quantitative often lead later on to qualitative differences.

Naturally a gene will, in general, have the more farreaching effects the earlier the stages of development it influences and the more numerous the processes it consequently affects. Therefore, the pleiotropic effects of genes often arise from the fact that they act in the early embryonic stages. For the same reason mutations having widespread somatic effects are frequently lethal. A relatively simple effect in an early embryonic stage can lead to numerous phenotypic changes. For instance, in a mutation of mice, produced by X-ray treatment by Little and Bragg, blood extravasates are caused by a recessive gene, and these bring in their train numerous and diverse abnormalities, such as clubfoot, polydactylism, anophthalmia, skull and brain defects, and so on.

The conception of heredity as a system of processes governed by harmonized reaction-velocities has farreaching phylogenetic consequences. De Beer [1] has shown that the limitations of Haeckel's fundamental

[1] G. R. de Beer, *Embryology and Evolution.* Oxford, 1926.

biogenetic law of the recapitulation of phylogeny in ontogeny depend to a great extent on the fact that, by way of changes of the rate of the developmental processes, the original sequence of ontogenetic stages, as it was present in the ancestors, can be mixed up in the descendants. Therefore we cannot speak of a " recapitulation " of the phylogenetic series in the ontogeny of the descendants, only of a repetition of ontogenetic states of the ancestors, the sequence of which may be profoundly altered.

Changes in the synchronization of developmental rates can, especially when they concern early embryonic processes, lead to far-reaching transformations. In mutations of Drosophila, such as *aristapedia* (leg instead of bristle), *tetraptera* (four-winged), *proboscidea* (transformation of parts of the mouth so that they become similar to the biting mouth-parts of other insect orders), the mutated gene does not lead to an isolated effect ; rather a quantitative change, simple in itself, controls a widespread process that leads to a profound transformation of the developmental pattern and hence to complicated morphological changes. In this sense Goldschmidt speaks of " hopeful monsters," the appearance of which can be decisive just for the great evolutionary transformations. From the palæontological standpoint, Schindewolf, in his theory of proterogenesis, has similarly stressed the importance of early embryonic transformations. It has already been mentioned (p. 77) that such changes are also important for the evolution of man. In certain cases a deeper analysis of the process of pattern-formation is possible on the principle of the harmonized reaction-velocities. Such case is the wing-pattern of butterflies (Goldschmidt, Henke, Kühn), which is especially suitable because of its two-dimensional nature and the possibility of developmental and genetical analysis. The formation of the manifold wing-patterns is based upon relatively few fundamental processes, which differ mainly quantitatively from one species to another and depend on a restricted number of Mendelian genes ; but they make possible numerous permutations and differences, thus producing the extraordinary wealth of design and colour in butter-flies. A second fundamental case is the harmonization of growth rates, a case where quantitative mathematical analysis is possible (p. 138).

What is a gene ? The total number of genes in Droso-phila can be assessed, on the one hand, from the number of mutations and crossovers found in experiment, on the other hand, from the number of chromomeres, repre-senting the site of genes, as found in microscopical obser-vation of the giant chromosomes. Both estimates lead to an order of eight or ten thousand genes in the four chromosomes of Drosophila. From this the volume of a gene can be calculated. It corresponds approximately to a small cube of about a hundred-thousandth of a milli-metre in length. In a similar way radiation genetics— the analysis of mutations caused by X-rays and other short waves (pp. 165 f.)—leads to the conclusion that the gene represents an atomic compound of the order of magnitude of a large molecule or a micella. A chromo-some, being a linear arrangement of genes, can be con-sidered as an " aperiodic crystal " (pp. 28, 31). A change in a gene can take place in mutations induced by radiation through the action of one light quantum (p. 166), and in spontaneous mutations probably through thermal motion. It indicates the sudden transition of the gene molecule into a new stable state, which can be interpreted as a transition into an isomeric form of the molecule, a change of side chains, or the like.

In 1937, the author has emphasized the necessity for an organismic conception of heredity :

" Chromosomes do not represent series of genes one of which produces perhaps vermilion eye-colour, the next miniature wings, others Bar eyes and short bristles, and so on. Rather the *whole* organism of the complete animal is produced from the *whole* genome of the germ cell. If we speak of ' genes,' this means nothing more than an expression of differences in the genomes located at certain points.

What we really find is only this : if we cross, say, a *vestigial* Drosophila with one having normal wings, analysis shows that there is a difference between the two races at a definite point on a definite chromosome. It is this difference that is localized at the point in question as the ' gene.' It is not a single gene, however, that produces this or that shape of wing. That wings (and all the numerous other organs) are formed at all is the work of the genome as a whole, except that the formation will be somewhat different according to which molecule is in the particular place on the chromosome. Genes are the expression of small differences in otherwise corresponding genomes. They are not points of origin for the individual organs. It may also be said that a chromosome-locus represents a sensible and easily disturbed point in the germplasm. It would be in keeping with this conception that mutations can be induced by injurious influences (X-rays, high temperature, etc.) and furthermore that the experimental mutations are to a great extent abnormalities. If the disturbance of the germplasm is especially strong it hinders development and acts as a ' lethal factor ' localized at a definite site. If it is weaker it produces malformations. With such a conception, many difficulties in the conception of the gene disappear, and we are no longer bothered by questions like that where the genes for higher systematic characters are localized, how they can be placed in chromosomes already filled with genes governing minor variations, and so on. On the other hand, the factual results of genetics remain unaltered."

These statements can be compared with the " Outlook on a future gene theory " given by Goldschmidt (*loc. cit.*). According to him, development in genetical research in recent years has led to a point where it is necessary to ask whether the conception of the gene as a unit of heredity with a separate existence is still tenable. Goldschmidt's considerations are based especially upon the position effect. In the mutation *Bar* of Drosophila, due to unequal crossing over, individuals occasionally appear having two neighbouring *Bar*-genes in their X chromosome. The effect of these *Bar*-factors, both lying in the same chromosome, is different from that of the two *Bar*-factors lying in the two X chromosomes of the female. The latter produce the normal *Bar* eye (with about sixty-eight ommatidia), but the former produce a much stronger reduction of the eye, known as the mutation *ultra-Bar* (with forty-five ommatidia). The effect of the *Bar* gene is thus dependent on its position. Furthermore, mutants that correspond phenotypically with gene mutations, that is, changes in a single gene, can appear as a result of translocations, inversions, duplications, and deficiencies, that is, as a result of a disturbance of the chromosome structure at or near the point of the gene concerned. These phenomena lead Goldschmidt to a theory of the germplasm " in which single genes no longer exist as separate units." The chromosome as a whole is a long molecular chain of a complex structure. Each point in the chain has a definite significance for the chemical properties and the effects of the whole. The Wild type is controlled by the whole chain as a unit. An alteration in the chain leads, however, to a disturbance of the catalytic reactions, which becomes manifest as a " mutation." A mutation may depend on an actual change at a definite place in the chromosome—and is then called a gene mutation ; it may also be due to a mere change in arrangement, as in the position effect. We can therefore describe the facts of genetics in terms of the gene concept ; however, the actual unit of heredity which controls development is the chromosome and the germplasm.

3. *Evolution I : The Tibetan Prayer-wheel*

The modern conception of evolution is based on genetics. In the essentials, it has come back to Darwin's theory of selection. An enormous overproduction of descendants takes place with all organisms. This does not, however, lead to an unrestricted increase in the number of individuals in a species, which remains more or less constant. There is therefore an incessant struggle for existence, which destroys the greater part of the individuals produced. On the other hand, in the life of the species small accidental variations from the normal—called mutations today—appear now and again. They may be unfavourable, neutral, or favourable. In the struggle for existence, natural selection eliminates the unfavourable mutations; the favourable ones, on the contrary, are preserved, and the organisms possessing them will more likely reproduce themselves. Through frequent repetition over long periods, this process leads to evolution, both as a creation of the multiplicity of forms in the world of the living and as a progressive adaptation of the organisms to their specific environments.

The first prerequisite of the theory of selection—the over-production of offspring, the fact that in spite of this no unlimited increase of the number of individuals takes place, and the resulting struggle for existence—is confirmed directly by observation.

The second prerequisite—the occurrence of inherited variations—is proved by modern research, since in every form sufficiently investigated mutations are discovered in abundance.

The third prerequisite—selection—is experimentally and mathematically proved. It is found, on the one hand, that different mutations can differ in their vigour and viability under different environmental conditions. On the other hand, a decisive means of testing the mechanism of selection is given in mathematical analysis, which, through the work of J. B. S. Haldane, Fisher, Sewall Wright, Ludwig, and others, is highly advanced in this field, though experiment and observation did not quite keep pace with the mathematical theory of selection. It shows that selection over relatively short periods allows favourable mutations, which at first appear in only small percentages, to become established. Taking an original mutation-rate of 1 per mille, the number of offspring as 25, and a selective advantage of 1 per mille (meaning that only 999 of the mutants perish, for every 1000 of the wild form) then only 710 generations will be needed for a dominant mutation, and 3460 for a recessive one to raise their frequency to 99 per cent of the total population. A certain difficulty exists in the fact that mutation rate is, in general, much smaller (approximately of the order of 10^{-6}). Therefore the " running-in-time," in which the frequency of mutants is raised perhaps from 10^{-6} to 10^{-3}, asks for a long period in the case of recessive mutations, and the overwhelming majority of mutations are recessive. Under the conditions assumed above, 100 million generations would be needed. Here auxiliary factors must be assumed, such as population waves, segregation into smaller inbreeding groups, and so on. These assumptions, however, correspond to factual biological conditions. On the other hand, experiment shows that many mutations have a much higher selective value than in the calculation given above.

To this must be added another important factor investigated in recent times, namely, the effect of isolation or the " drift " principle, studied especially by Sewall Wright. If a species is subdivided into small populations isolated from each other, then the mere accident of gene-combinations can lead to the establishment of different mutations in these populations, irrespective of their selective value; and this may lead to the splitting-up of the originally uniform species into different subspecies and eventually species.

The title of Darwin's work has to some extent concealed the fact that the " origin of species " is only one of the problems of evolution and not the greatest. Some four main problems can be distinguished : first, the origin of the multiplicity of forms within a given type of organization or *bauplan*, that is, the origin of the lower systematic units, of races, subspecies, species, perhaps also of genera; secondly, the origin of these types of organization themselves, that is, of the higher systematic units; thirdly, the origin of ecological adaptations to definite environments; fourthly, the origin of the complex morphological and physiological integration within the organism as a whole. It is, of course, impossible to draw sharp border lines between these problems. Problems 1 and 2 concern the origin of the multiplicity of organic forms, 3 and 4 that of the "fitness" of the organism.

It is generally accepted that the formation of subspecies, and most probably of species as well, is the result of the factors already mentioned, which are established experimentally and theoretically. These factors are random mutation, selection, and chance action within small isolated populations. This field of research offers many interesting problems, but there are hardly basic difficulties or controversies. Similarly, the theory of selection provides a satisfactory explanation for many adaptations, such as protective resemblance and mimicry, in which an animal like the Dead Leaf Butterfly imitates part of a plant, or a defenceless animal mimics an inedible one, as is often the case with butterflies. It is to be assumed that in the beginning chance variations occurred that caused a slight resemblance to the protected form, and thus gave an advantage in selection; later on, this resemblance was enhanced by natural selection, and led finally to a complete imitation. So there is hardly any question that the modern theory of selection explains problems 1 and 3, and hence what is called micro-evolution. No complete agreement is reached with respect to questions 2 and 4, the so-called macro-evolution. The overwhelming majority of modern biologists, and especially the investigators connected with genetics, accept the theory of selection. This attitude is based, empirically, on the success of genetics, as such and in its application to evolution; methodologically it is based on the principle that only factors which are known and exactly and experimentally demonstrable should be admitted. Some morphologists and palæontologists, however, hold opposite opinions. Workers in these fields, more so than experimenters, are incessantly confronted with the marvellous architecture of living organization, its integration,

G

and the correspondence of structure and function, and are reluctant to believe that these are a product of mere chance. Similarly, the physiologist looks at the amazing complexity of the system of catalysers in a cell wherein the loss of even one member may make it degenerate into a cancer-cell, or surveys the number of conditions necessary for the normal function of a gland, or of the nervous connections necessary even for a simple reflex; he feels somewhat uncomfortable to accept the explanation that this is all due to chance.

As already mentioned, the mechanisms assumed— mutation, selection, isolation—are experimentally verified. However, apart from some cases occurring in polyploid plants, no new species has ever arisen within the sphere of observation, let alone " macro-evolutionary " changes. Selection Theory is an extrapolation, the boldness of which is made acceptable by the impressiveness of its basic conception. With a less picturesque theory, one would doubtless hesitate to extend cosmically and universally a principle which is controlled experimentally only to a rather limited extent. The pros and cons of the theory of selection have been discussed on countless occasions. Indeed, the controversy with more or less noteworthy " objections," constitutes a main part of every presentation of the theory—a procedure that would be looked for in vain in a treatise on, say, physics or physiology. It is a methodological maxim, which is unreservedly correct, to take into account only known and experimentally demonstrable factors, to extend according to the actuality principle, their application as far as possible, and to exclude, in the sense of Occam's razor, factors unknown and experimentally unproved. On the other hand, for not quite fifty years we have carried out genetical research on some dozens of animal and plant species, the mutations of which never transgressed the limits of the species. It is a bold extrapolation that nothing else has happened in a billion years or so of evolution " from amœba to man." So the dispute concerns ways of thinking rather than factual evidence.

Innumerable characters of the organisms are seemingly useless. To a great extent they are precisely those that the taxonomist considers most important, because on account of their functional neutrality they show a high degree of constancy. The multiplicity of leaf-forms, the arrangement of leaves, the numbers three, four, or five in flowers, the number seven characteristic of the cervical vertebræ in mammals, which is found in the neckless whale as well as in the long-necked giraffe; all such morphological peculiarities which form the framework of taxonomy seem to represent "types" that in themselves have nothing to do with usefulness but can be fitted to different exigencies and habitats much in the same way as churches, town halls, or castles can be built equally well in gothic, baroque, or rococo styles. Goebel, the great botanist, has emphasized that the manifoldness of organic forms is much greater than the manifoldness of environmental conditions. In the same part of the sea, in a thoroughly uniform environment, hundreds of species of foraminifera or radiolaria can be found; "natural art-forms," the fantastic diversity of shapes of which has obviously nothing to do with usefulness. The selectionist, however, is not embarrassed by these arguments. Structures which are useless in themselves can be preserved in uniform surroundings where selection-pressure is low. They may arise by chance, according to the Sewall-Wright principle, in a species subdivided into small isolated populations; or useless characters were linked with others having a selective advantage, perhaps mere differences in viability, and thus were perpetuated.

In the living world the maxims " Why make it simple, when it can be complicated ? " and " It can be done this way, but it can also be done the other way " seem to be much in vogue. Often astonishingly roundabout ways are taken to reach a goal that could be reached far more simply and with less risk. Consider the life-cycles of many parasites, the relatively simple one of the human Ascaris, for example. The young larvæ enter the intestine with the food; they pass through the intestinal wall into the blood and into the circulatory system, through the portal vein into the liver, then into the lung and throat, where they must be again swallowed, to arrive back, at the stage of sexual maturity, in the intestine, where, in our humble opinion, they could quite well have remained. In the remarkable trap flowers, such as Lady's Slipper or the wild Arum, the insects that help to pollinate the flowers are captured and imprisoned by intricate devices so that they can be used for pollination in due time. The net outcome of these intricate mechanisms, however, is that in our climates the Lady's Slipper and Arum are among the more rare plants almost on the verge of extinction. Even in their tropical centres, the families to which they belong do no better than plants utilizing simple wind-pollination. " It can be done this way, but it can also be done the other way." The ruminants have acquired an extraordinarily complicated multiple stomach, doubtless an organ of the greatest usefulness and highest selective value for animals digesting vegetable food ; but horses have a simple stomach, and attain the same remarkable body size and geographical distribution. With an artist's industry, nature paints in mimicry on the wings of a butterfly the copy of a protected model in order to make the birds believe that the imitation also tastes bad. But the ordinary Black Archers, although readily taken by birds, are as numberless as the sands of the seashore, and strip the forests bare. The same survival in the struggle for existence, retorts the selectionist, can be attained by different means, and the very existence of all these forms shows their fitness.

The same holds for absurd and even apparently dis-advantageous formations, such as the giant and baroque forms of the ammonites, the monstrous horns of the titanotheres, the antlers of the megaceros, which, being heavy and hindering a forest animal in its movements, pre-sumably accounted for its extinction. Ludwig has stated that selectionism offers some fourteen to twenty possible ways of explaining such disadvantageous characters. This shows that they are not a refutation of the theory of selection, but equally that a decisive judgment is impossible. For one hypothesis that can be neatly proved or disproved has much greater value than a host of possibilities. To mention some of these explanations : a character that is disadvantageous or neutral at present may have been advantageous in the past; a disadvantageous character may be correlated by pleiotropism with one having selective value, as in titanotheres the apparently useless horn formation is correlated, according to the law of allometry (p. 138), with the advantageous increase in body size; the character in question owes its origin to sexual selection; within a species the existence of which is secured interspecifically, intraspecific selection may take place, which can lead eventually to developments harmful to the species itself; and so on.

A lover of paradox could say that the main objection to selection theory is that it cannot be disproved. With a good theory, it must be possible to indicate an *experimentum crucis*, the negative outcome of which would refute it. If the gravitational force exerted by the planets was proportional to $1/r^3$ instead of $1/r^2$, Newtonian mechanics would be wrong; if a cuneiform text of logarithmic tables was found somewhere, we should have to revise our notions on Babylonian mathematics. In the case of Selection Theory, however, it appears impossible to indicate any biological phenomenon that would plainly refute it.

Consider the construction of an accommodating eye ; a soft lens, a ciliary body, ciliary muscles, nerves leading to and from corresponding centres must all be present and work together to allow its functioning. There is a big difference between the " characters " that are studied in experimental genetics, and which represent small variations of, for instance, the size of an organ, its colour, and the like, and the origin of " systems " which are useful and have survival value only as organized wholes. In this sense, single and accidental mutations cannot lead to a gradual development or improvement of the apparatus, but can only spoil it ; the absence of one single part makes the whole system into a sort of useless or even deleterious tumour. Thus it is calculated how infinitely improbable it is that such co-adaptation, the origin of apparatus capable of functioning only as a whole, resulted from accidental mutations. The selectionist's answer: Remember that the lens-eye is the product of an enormously long evolution. Imagine sufficiently minute intermediate stages from a simple pigment-spot through the socket-eye up to the lens-eye, stages as they are actually shown by comparative anatomy, each one giving a small selective advantage. Then you will understand how such formations have been added up in the long time of phylogenetic history. And the selectionist also gives nice calculations that support his view.

With the same stubbornness the uniformity of micro-evolution and macro-evolution, of the origin of the multiplicity of forms within a " type " and the origin of these types themselves, is challenged and defended. The rise, say, of a winged insect from unwinged ancestors is on a different level from the known mutations in Drosophila, which concern only differences of formation in already existing wings. The macro-evolutionary origin of a " type " is not the product of a gradual accumulation of small changes, but of " macro-mutations " that give rise to far-reaching transformations in the early embryonic stages. This view is supported by palæontology, which indicates two phases in evolution; a first one, the sudden rise of a new type, which, immediately after its appearance, splits up, explosively as it were, into the main classes or orders; a second one, the slowly progressing speciation and adaptation to different environmental conditions within those groups. The selectionist's answer is: a clear definition of what shall be called " type " is not possible, so a boundary cannot be drawn between " macro-evolution " and " micro-evolution." A number of profoundly transforming mutations are known in experiment, such as the four-winged mutation *tetraptera* of the dipter Drosophila or mutations of the snapdragon with radial instead of bilateral symmetry of the corolla.

The fact that transition stages between different " types " are rare or often missing is easily explained, for new types go back to thin ancestral roots, and accordingly the probability of the preservation of these ancestors as fossils is small. Nevertheless, we have such transition stages in good measure, for example, in *Archæopteryx* from reptile to bird, or in the almost unbroken series that leads from reptiles through the *Theriodontia* to mammals. Indeed, there is no reason to assume a basic distinction between micro-evolution and macro-evolution, or that the laws of heredity were in the past different from those of today.

It is often asserted that evolutionary progress cannot be understood in terms of " usefulness." If higher organization means selective advantage, the higher organisms should have supplanted the lower ones. Every cross-section of nature, however, shows the most diverse levels of organization from unicellulars up to vertebrates, all maintaining themselves perfectly, and indeed all necessary for the maintenance of the biocœnosis. The selectionists retort : When man invented the bow and arrow the simpler method of clubbing became obsolete ; the introduction of firearms condemned the armoured knights to extinction ; tanks make cavalry attacks of doubtful value. In the present struggle for existence among nations, survival is granted only by aircraft— until in the near future perfected atomic bombs will save mankind from bothering further about questions of selection, both with respect to theory and to their own survival. In this wonderful progress earlier stages may well be preserved. The backwoodsman may persist on the level of civilization of the Merovingians, and less · efficient methods of killing each other can be preserved in the hinterlands of Central Africa or New Guinea. Similarly, the sluggish saurians were replaced by the more versatile, warm-blooded mammals, the marsupials by the *Placentalia*, but this does not alter the fact that lizards, snakes, tortoises, and crocodiles have survived to remind us of the former reptilian splendour, or that marsupials still exist in Australia, where no higher mammals had come in.

Discussions of this sort can be continued to mutual exhaustion, but do not, however, lead to the conviction of the respective opponents. The reason is understandable. We have one or two dozen experiments in which the " usefulness " of a character is experimentally demonstrated, as when, for example, those individuals of a species of insect which have a colour identical with the background are less eaten by birds than those with contrasting colour. But for the extrapolation that evolution was controlled by " usefulness " there is no way of experimental verification or falsification. If any species has survived and has undergone further evolution, then the change must have been either advantageous or related to advantageous changes or at least not deleterious, for otherwise the species would have simply been extinguished. But this ever remains a *vaticinatio post eventum*. Like a Tibetan prayer-wheel, Selection Theory murmurs untiringly : "Everything is useful." But as to what actually happened and which lines evolution has actually followed, selection theory says nothing, for the evolution is the product of " chance," and therein obeys no " law."

But is this correct ?

4. *Evolution II : Chance and Law*

Is evolution a process accidental in itself and directed only through outside factors, namely, is it a product of random mutation and equally accidental environmental conditions resulting in the struggle for existence and selection, plus the accidental effect of isolation and subsequent speciation ? Or is evolution determined or co-determined by laws lying in the organisms themselves ?

This is a question that leads us out of the tug-of-war of opinions, different interpretations and hypotheses, and which can be judged on the basis of facts.

We must start from a fundamental statement. Mathematical analysis shows that selection-pressure is greatly superior to mutation-pressure : even a small selective advantage in a positive or negative way is much more effective than directed mutation without selection, even if the mutation in question should appear at a high rate and repeatedly. Therefore " directiveness " of evolution in the sense that it works against selection would be impossible; in the sense that it works without selection, it would be effective only over exceedingly long periods of time.

From these statements and the " undirectiveness " of mutations (p. 94), selectionism concludes that the direction of evolution is determined only by external factors. But this conclusion does not follow from the premises. If selection represents a *necessary* condition of evolution, it does not follow that it indicates a *sufficient* condition.

Perhaps a physical analogy can make this clear. The principle of entropy holds for all (macro-) physical processes. It indicates a limiting condition, which—apart from the exception of heat movements of molecules in very small spaces—prescribes that in every physical process a certain quantity, called " entropy," increases. But the entropy principle defines only the general direction of events. In general, many processes are thermodynamically allowed; but whether anything happens at all, and if so what really happens, we cannot learn from the entropy principle alone, but we must take into account the specific conditions of the system. Whether, say, a thermodynamically possible oxidation actually occurs, or why alum crystallizes in octahedras but Iceland spar in rhombohedras, we are not told by the principle of entropy though it is obeyed in oxidation as well as in crystallization. For such knowledge we must be informed about the nature of the reacting substances, their rates of reaction, the lattice forces of the different kinds of molecules, and the like. In a similar way the principle of selection indicates a limiting condition, which—apart from exceptional cases where selection-pressure is absent —prescribes that in every evolutionary process the " advantage " for the organism in question increases. But if and what happens in the individual case cannot be inferred from the principle of selection. For instance, nothing at all may happen, as in the case of the sparrow, which was accidentally introduced into America but produced no new races there, or in the Brachiopod *Lingula*, which has remained for some hundreds of millions of years unchanged. Or the same effect—namely, preservation in the struggle for existence—may be achieved in quite different ways. Both thermodynamics and the selection principle are the consequences of a mechanism of chance. The biological assertion, that everything is said with the latter, may be compared to the " energetics " of Ostwald, now out of date, which believed that physics is all contained in the energy principles.

If now we examine the mutations in Drosophila, for example, we find that they give the impression of an uncontrolled multiplicity of variations. Furthermore, spontaneous mutations, as well as those induced by outside factors, are " accidental " with respect to external conditions, that is, they show no adaptive character. It is not the case that mutations which appear, say, at increased temperature represent adaptations to that higher temperature. Only the rate of mutations which also appear otherwise is increased. However, the multiplicity of mutations and their lack of adaptiveness and direction with respect to external influences do not necessarily mean that they are entirely fortuitous. Rather there are indications that mutations and evolutionary changes have many, but not an infinite number, of degrees of freedom.

Of course, mutation is limited, in the first place, by the nature of the genes present and the possibilities of their variation. Putting it crudely, a vertebrate will never produce a mutation that leads to the formation of a chitinous cover such as is found in insects, because this is not within the range of a vertebrate's anatomical and physiological ground-plan. The same holds for finer details. For instance, green colour is found rather seldom among butterflies, though it is common at the caterpillar stage and is, moreover, an excellent protective colour. Blue roses and black tulips could not be produced in spite of the florists' efforts over many centuries.

Investigation into the nature of the gene and of mutation leads to a similar conclusion. Since a gene is a physico-chemical structural unit of the nature of a large protein molecule, and since a mutation represents a transition to a new stable state by way of isomerization or a change in side chains or the like, a change will certainly be possible in a number of directions but not in all directions, in a similar way as only certain quantum states are permitted to an atom. In both cases, quantization is at the basis of the jump-like character of the change, as well as of the high stability and organization of the system. An atom cannot take up any small quantities of energy it is exposed to under the constant bombardment of the heat movements of the surrounding particles; only a complete quantum jump will induce a change in it. This ensures that it may remain unchanged over an indefinite time. In the same way, the " quantized " character of mutations is at the basis, first, of their discontinuity, secondly, of the great stability of the gene and the relative rarity of mutations, and thirdly, it follows that the number of possible mutations is not infinite, since only certain stable states are " allowed."

Here we come to an important problem. The theory of evolution, based upon an enormous amount of factual evidence, states that the animal and plant kingdoms have arisen, in the course of geological time, from simpler and more primitive forms to more complicated and more highly organized ones. Genetical experience leads us to accept as a fact that this has happened by way of step-like mutations. Actually, however, we find no evidence either in the living world of today or of past geological epochs for a continuous transition. What we actually find are separate and well-distinguished species. Even the existence of more or less numerous mutations, races, subspecies, etc., within the species does not alter the basic fact that intermediate stages from one species to another which should be found if there were a gradual transition, are not met with. The worlds of organisms, living and extinct, do not represent a continuum but a discontinuum.

The discontinuity of species is based presumably on the fact that certain conditions of stability exist not only for the individual genes but also for genomes. In individual genes the transition from one state to another is discontinuous, and this is the reason for the jump-like character of mutation. As regards the conditions of stability for genomes, the following conception may be stated. A " species " represents a state in which a harmoniously stabilized " genic balance " has been established, that is, a state in which the genes are internally so adapted to each other that an undisturbed and harmonious course of development is guaranteed. If there are no external disturbances, stability is ensured for a theoretically unlimited number of generations. If a mutation occurs it means a disturbance of this pattern ; therefore in the majority of cases, mutations will lead to unfavourable, even lethal, results, even before selection is taken into account. But every gene acts not only in its own right but more or less also as a modifying factor influencing the action of the rest of the genome (p. 74). Species always differ from one another in a great number of genes. The more numerous the mutations that have occurred, however, the greater is the chance of a disturbance of the established genic balance, even if these mutations in themselves would be favourable. Hence a form which is on the way from one species to another will be in a state of instability, and it will be particularly exposed to selection. So this transition must be passed through quickly, or speaking statistically, such transitional stages will be comparatively rare. Finally, the form, provided it does not in the meantime become extinct, reaches a new state of genic balance in which it can remain again for a long time. This gives precisely the state of affairs we find in nature, namely, in general, well-defined species showing mutations but not continuous inter-gradation, and only in exceptional cases *formenkreise*, or groups of intergrading subspecies which seem to be just in the process of speciation.

A similar argument probably holds also for the origin

of the great types of organization. It was remarked earlier (pp. 90 f.) that no exact borderline between micro-evolution and macro-evolution can be drawn. Nevertheless, presupposing that the process of evolution is continuous, just in the transition from one type to another, long periods of time and corresponding numbers of intermediate stages should be expected. But we do not find this; instead an unknown X frequently stands just on the decisive branching point where a new type has arisen. And if intermediates are known, such as *Archæopteryx*, the link between reptiles and birds, or *Peripatus* between Annelida and Tracheata, or the theriodont series of the South African Karroo, then these forms are rare or restricted to formations of fairly short duration. *Archæopteryx*, for example, is represented by only two specimens, as compared to the many thousands of reptilian fossils found in the slates of Solenhofen. For such reasons Schindewolf and other palæontologists contend that evolution is not a continuous process but shows a periodization: first, an explosive phase of type formation, with a quick splitting into the main groups; second, a phase of slower speciation and progressive adaptation to different habitats within each separate group; and finally, a phase of decadence in which the multiplicity of forms runs wild, so to speak, and which leads in the end to extinction. Schindewolf is probably right in emphasizing that the usual method of demonstrating the sufficiency of geological time for a continuous transformation by way of small mutations and selection is incorrect; only the beginning and the end of the evolutionary process is taken into account, and the intervening period is assumed to be filled with uniformly distributed, gradual transformations. Actually, however, the multiplicity of forms appears suddenly. Within a series of species, a new one appears, and instead of proceeding continuously to the next, remains constant for hundreds of thousands of years, only then to give rise to its successor. Within types, the great classes are present from the beginning. For example, the main classes of Angiosperms are already present in the lower Cretaceous, and similarly the main orders of the Placentalia, such as insectivores, rodents, carnivores, ungulates, and primates, at the opening of the Tertiary. The long periods of time that follow are filled out with only minor transformations of these basic types. This phenomenon is one of the foundations of Schindewolf's theory of proterogenesis, that is, of sudden early ontogenetic transformations which lead to the origin of new types. This view is similar to Goldschmidt's theory of hopeful monsters (p. 79). Assuming the correctness of the periodization mentioned, there seems, however, no need to assume a somewhat arbitrary gulf between micro-evolution and macro-evolution, or a unique character of the supposed "macromutations." The relatively small number of fundamental types of organization in the animal and plant kingdoms demonstrates that the great evolutionary transformations are based upon relatively rare and profound genetic changes; this does not imply, however, as is stated also by Schindewolf, that they are basically different from the known kinds of mutation. The sudden appearance of types and the lack of intermediates can again be explained by considering that the transitions between types, like those between species, represented unstable and hence short-lived states.

With the question of the conditions of stability, which apply to the preservation of species and types as well as to the transition from one state to another, we have claimed for a " Statics," so to speak, of evolution. These questions can be answered today only in a preliminary way. The other main complex of problems is a " Dynamics of evolution," that is, the laws governing evolutionary changes. In this field there are some promising beginnings.

Every law of nature implies a statement of recurrences. If, therefore, we wish to establish laws of evolution we must look for such recurrences. We find them in the phenomenon that parallel changes appear in many groups of organisms. Three kinds of such parallelism can be distinguished, though it is, of course, often difficult to keep them apart in an individual case. The first kind comprises changes in homologous genes; the second, parallel changes in the processes of development, that is, similar phenotypes that are due to different genic or environmental factors; the third comprises parallelisms arising on a different genetic and developmental basis.

Many genes, and the mutations due to changes in them, are demonstrably homologous. Such homology has been demonstrated, for instance, for extensive chromosome regions in different species of Drosophila. It is especially illustrative when the genes concerned influence development by way of known gene-controlled substances or so-called gene-hormones. For example, the a^+-substance, which in the flour moth (*Ephestia*) leads to pigmentation, is identical with a v^+-substance, which makes v- (vermilion) eyes in Drosophila take on the normal dark red of the Wild type. Like hormones proper, these gene-controlled substances are not species-specific; v^+- and a^+-substances are homologous gene-hormones, v (vermilion eyes in Drosophila) and a (absence of pigmentation in the flour-moth), and correspondingly v^+ (Wild-type eye colour in Drosophila) and a^+ (Wild type of the flour moth) are homologous mutations and genes respectively. Parallel mutations have considerable clinical interest, since corresponding hereditary diseases are found in man and domestic animals which are caused by parallel mutations and are, in animals, accessible to genetical analysis. The interesting consequence is that organisms which are widely different taxonomically, such as the flour moth and the fruit fly, rabbit and man, have certain genes in common, and hence show parallel mutations. On the other hand, it appears that, to a great extent, it is not differences in individual genes but the gene-complex, that is, the co-ordination of the genes as a whole, which is specific for a taxonomic group. The bearing of this principle will be examined later (Vol. II) in an important field, that of morphological forms. We shall see that a considerable part of evolutionary changes and of the multiplicity of living forms depends on changes in proportion, which in their turn depend on differences in the harmonization of growth rates and hence on differences in co-ordination within the gene-complex.

Parallel mutations within smaller or larger groups are a common phenomenon. For example, we find bearded and non-bearded forms, brittle and firm ears, summer and winter forms, and so on, in different species of wheat; similarly, the genus " rye " repeats in detail the series of species which is found within the genus " wheat." This " law of homologous series " (Vaviloff) in different systematic groups is not without practical significance. In cultivated plants where a desired mutation is apparently absent, but is found in related species and genera, it may be expected that it can probably be discovered by further intensive investigation, or can be produced artificially by means of radiation.

A second class is represented by phenotypically corresponding mutations due to non-homologous genes. We find albino forms, for example, in widely different species, such as rabbits, mice, cats, humans, etc. Some of these albinos are due to mutations of homologous genes; others certainly to mutations of non-homologous genes, for albinism always appears when one of the factors (often numerous) necessary to develop pigmentation is missing. Obviously this phenomenon indicates that, as mentioned earlier (p. 74), the genes are influencing the complicated physiological process that leads from the genome to the completed organism. Different factors may affect this process in a similar way, and phenotypically similar variations can arise as hereditary mutations of different mutated genes, and even as non-inherited phenocopies due to environmental factors.

It is not easy to decide in a particular case whether a given parallelism belongs to the first or the second class, that is, whether it is based on genic homology or parallel deviation of developmental processes without such homology. In any case, we meet parallelisms nearly everywhere in the fields of botany, zoology, anthropology, palæontology, and zoogeography, although no comprehensive treatment seems to exist as yet. For example, breeding and domestication provide an ample field. In the most different species, we find parallel variations of the coat, such as albinism, melanism, piebalding, striping, curling, and so on ; we find variations in the skull, such as " pug head " and dolichocephaly ; " dachshund legs," and other variations, which can with a fair degree of probability be attributed partly to mutations of homologous genes and which reveal a common inheritance of taxonomically widely different species. In man certain mutations, such as, for example, the kinky-hair factor, and dwarfism in the different pigmy races, have probably appeared repeatedly in all the principal races, independently of each other in space and time. A similar thing holds for the types of constitution in man that are present in all races, and with which even constitutional types of domestic animals can be compared. Finally, in palæontology, parallel series are widespread. They are found in the different independent groups of ammonites, titanotheres, and so on. Evolutionary trends such as that in titanotheres, are determined by a quantitative law, that of allometry (p. 138). Similarly, in zoogeography the so-called Geographical Rules, such as Bergmann's rule of the greater body size of animals in colder as compared with those in warmer climates, Allen's rule of the shortening of exposed portions of the body (limbs, tails, ears) in cooler regions, and Gloger's rule of the decrease of melanin pigmentation with mean temperature and aridity, are based, at least partly, on parallel evolution.

Thirdly, parallelisms can appear when the genetic as well as the developmental basis is different. For example, it is fairly probable that the phenomenon of mimicry is partly based on genic homology. In some butterflies, however, an excellent mimicry pattern is produced by means of pigments which are chemically quite different from those in the model. The similarity here exists in spite of a different genetical basis, and must have originated through selection.

Among the parallelisms which are based neither on

H

genic nor developmental correspondence, we find first those we can term ecological parallelisms. To these belong the analogies according to the classical definition, namely, organs similar in function but differing with respect to their position in the *bauplan* and in their phylogenetic origin, such as the wings of birds and insects, or the fœtal membranes in mammals, insects, and other animals. The term "convergence" should be limited to parallel formations that have arisen, as adaptations to similar environmental conditions, independently in different groups of a common phylogenetic origin. In other words, we call "convergent" such organs and structures which are homologous, developed later on in different directions, and finally became analogous. An example is the streamlined body shape and the fins as adaptations to an aquatic life, which are primary in fish but secondary—*via* terrestrial ancestors and return to the sea—in ichthyosaurians and whales. Other examples are the similar adaptative types in marsupials and placental mammals, the succulent stems of desert plants belonging to the different families cacti, *Euphorbiaceæ*, and *Asclepiadaceæ*. Such ecological parallelisms easily fit into the usual scheme of adaptation and of the evolutionary factors producing it.

However, there are doubtless evolutionary principles that allow only certain trends in the evolution of many organs. To cite the most famous example, lens-eyes constructed on the principle of the camera are found in very similar form in the scallop, the cuttle-fish, and vertebrates. Phylogenetically these forms are widely diverse, and the development of their eyes differs ontogenetically. In invertebrates the eye is derived from the epidermis, in vertebrates it is a derivative of the brain. Anyway, once the way towards formation of a complicated eye is taken in phylogeny, nature obviously has no other course than to pass through the successive stages of flat-eye, socket-eye, and lens-eye, and thus we find them in the most divergent classes of animals. The same applies to fundamental principles of organization. For example, we find a parallel evolution of a secondary body cavity (cœlom), of metamerism, of a circulatory system, in the phyla of Protostomia and Deuterostomia, in the annelids and in the chordates; phyla that are antithetic in their *bauplan*, their phylogeny, and ontogeny. The same thing applies to physiological characters. For instance, there exists only a small number of respiratory pigments in the animal kingdom, mainly hæmoglobin, chlorocruorin, hæmerythrin, and hæmocyanin; they are found independently in different animal groups, such as in vertebrates, ram's-horn snails (*Planorbis*), and blood-worms (larvæ of the gnat *Chironomus*). The number of possible proteins has been estimated at 10^{2700}, contrasted with a total number of 10^{79} electrons in the universe. But although the formation of respiratory pigments has happened a number of times (probably from oxidative enzymes, which are common in most organisms), it could take only one of these few directions.

Thus the changes undergone by organisms in the course of evolution do not appear to be completely fortuitous and accidental; rather they are restricted, first by the variations possible in the genes, secondly, by those possible in development, that is, in the action of the genic system, thirdly, by general laws of organization.

These factors seem to be responsible for the fact that evolution often conveys the impression of " orthogenesis," that is, of inherent trends in definite directions. As already stated, mathematical analysis shows that the effect of selection-pressure is much greater than that of mutation-pressure within the known rates of mutation. Therefore, orthogenesis, in the sense of trends to determine the evolutionary process in the teeth of selection, will be an exceptional phenomenon, or perhaps not exist at all. But there is orthogenesis in the sense that evolution is not determined merely by accidental factors of the environment and the resulting struggle for existence, but also by internal factors. " Blind alleys in evolution," that is, evolution in an unfavourable direction, seems to appear where selection slackens down, as when a group of organisms has reached unchallenged domination. Then those " excessive for-mations " may appear which usually herald imminent extinction. We find them, for example, in the giant reptiles and the ammonites at the end of the Mesozoic, similarly in the titanotheres (p. 89), the megaceros, etc. The conditions are similar to those of domestication, in which a multiplicity of partly monstrous forms is pre-served in a protected species, forms which would be rapidly eliminated in wild life. Remember, for example, the great number of varieties of domestic dogs and pigeons, the many albino races, the waltzing mice showing an inherited disturbance of the inner ear; similarly, " phenomena of domestication " such as bulldog types, caries, and the like have appeared in the cave bear towards the end of the glacial period. Again, the same principle leads, in man, to the increase and spreading of variations, such as short-sightedness, tooth decay, here-ditary predispositions to disease, which would be weeded out speedily in a wild state but which, under conditions of civilization, are no longer a menace to survival.

In this sense, orthogenesis may lead to favourable results, as was the case, for example, with the increasing size of brain in the series of mammals. But it can also lead, as in the examples given, to blind alleys in evolution. Within this orthogenetic evolution, however, a " prin-ciple of utilization " is active. It is not progressive adaptation that brings forth orthogenesis; but ortho-genetic trends may eventually create prerequisites for new and higher achievements. For example, different species of apes show different degrees of cephalization, which, after Dubois' theory, signifies a doubling of brain size from one group to the next higher one; but their habitat and behaviour are not much different. Similarly, primitive races in man have the same cephalization as civilized man; but this highly developed brain is cer-tainly not fully utilized in the primitives, and perhaps not even in the present state of civilization.

While fully appreciating modern selection theory,

nevertheless we arrive at an essentially different view of evolution. It appears to be not a series of accidents, the course of which is determined only by the change of environments during earth's history and the resulting struggle for existence, which leads to selection within a chaotic material of mutations. No more is it, to be sure, the work of mysterious factors, such as a striving for perfection or a tendency towards purposiveness or adaptation. But it appears as a process essentially co-determined by organic laws, which in suitable cases can be formulated in an exact way. We need not commit ourselves with respect to the factors of evolution. We may leave open the question whether evolution has followed laws because the mutations that have occurred have had certain lines of preference or only because " orthoselection " has favoured certain evolutionary trends; whether or not " macromutations " have occurred in phylogeny which were essentially different from the mutations known in experiment. This we cannot decide, because we cannot reproduce evolution, and it is also of secondary importance. What really matters is the statement that evolution is not a process at random, but is governed by definite laws, and we believe that the discovery of these laws constitutes one of the most important tasks of the future theory of evolution.[1]

[1] After this book was written the author became acquainted with the excellent work by B. Rensch, *Neuere Probleme der Abstammungslehre*. Stuttgart, 1947. Rensch's fundamental question is whether the factors that are known to be basic in speciation (mutation, selection, fluctuations of populations, isolation) are sufficient also to explain macroevolution (transspecific evolution, in Rensch's terminology); or whether specific autonomous evolutionary forces are to be assumed for the latter. Rensch answers in the first alternative. However, neither the organization of the animal body nor the environmental changes leave room for completely accidental evolutionary changes; there exist limiting conditions which, in many instances, are operative as an " evolutionary constraint." Rensch's book is probably the first attempt to integrate the laws of macro-evolution. The author gladly acknowledges the agreement with his own views, as stated especially in his *Theoretische Biologie*, Vol. II, and emphatically refers the reader to Rensch's book. We shall come back to these problems in the following volume, when the problems of " Dynamic Morphology " will be presented in detail.

5. *Evolution III : Unscientific Interlude*

The first Darwinist to explain organic "fitness" on the basis of events at random was the pre-Socratic Empedocles. In a famous fragment it is said that life sprang from the moist earth under the influence of fire; first single heads, eyes, limbs were formed, which united into monsters, such as creatures with many hands, cattle with men's heads, human bodies with bull's heads, until creatures capable of survival also appeared accidentally; these were the ancestors of the plants and animals of today. It is also known that Darwin's theory was an application of contemporary national economy to biological science. A most important starting point for Darwin was Malthus's statement that living beings multiply at a higher rate than the quantity of food available increases. Similarly, the consideration of biological phenomena in terms of "profit" and "competition" corresponded to the national economy of the Manchester school. In these general conceptions the emotional antipathy against "Darwinism" is rooted. On the one hand, there is the paradox wittily stated by Ludwig that "the evolution from the ancestral worm up to Goethe and Beethoven appears to be a product of lucky 'accidents' which happened to individual genes." On the other hand, here is the root of Nietzsche's resentment against "the pedlar's philosophy" of Darwinism with its "smell of poverty" of masses that are struggling for their brutish existence.

In the theory of evolution the same alternative exists which we find in all fields of biology. On the one hand, there is the mechanistic theory, considering life as purposeless and accidental, because this appears to be the only basis for a truly scientific theory. On the other hand, there is the opposition against this view, and the only alternative seems to be to assume factors which are scientifically uncontrollable and mythical. The synthesis here, as elsewhere, is in the organismic law.

For naïve and unbiased contemplation nature does not

look like a calculating merchant; rather she looks like a whimsical artist, creative out of an exuberant fantasy and destroying her own work in romantic irony. The principles of " economy " and of " fitness " are true only in a Pickwickian sense. On the one hand, nature is a niggard when she insists on abolishing, say, an already minute rudimentary organ; this little economy having, as maintained by the theory of selection, enough advantage to be decisive in the struggle for existence. On the other hand, she produces a wealth of colour, form, and other creations, which, as far as we can see, is completely useless. Consider, for example, the exquisite artistry of butterflies' wings, which has nothing to do with function, and cannot even be appreciated by their bearers with their imperfect eyes.

This productivity and joy of creation seems to be expressed in the " horizontal " multiplicity of forms on the same level of organization (p. 87) as well as in the " vertical " progress of organization, which can, but need not necessarily, be considered as " useful " (p. 91). We have already said that such considerations are not a refutation of the theory of selection. There are, however, two aspects where we remain dissatisfied.

From the viewpoint of science, we are not satisfied with the meagre answer that all this is possible within the range of the established factors of evolution, that it has arisen somehow by a selection of the advantageous or been allowed to survive merely by the accidents of gene distribution of indifferent mutations. We rather want to know the " secret law " at which " the chorus hints."

On the other hand, being people of our time, we are inclined to consider the utilitarian theory as a sort of living fossil like the tuatara of New Zealand, as a relic of Victorian Upper-Middle class philosophy. It is the projection of the sociological situation of the nineteenth and early twentieth centuries into two billion years of the earth's history. " Progress " is " useful," therefore, " usefulness " is the " cause of progress." A later " great time " converted this idea, without actually

changing its meaning, into the notion of struggle as the
father of all things.   But the sociological parallel upon
which Darwinism is based is not convincing any more.
The major term of that deduction is trivial or frivolous,
the minor term certainly false.   The progress of mankind
in science and technology has certainly not arisen from
the need for improved adaptation.   First, there was the
theory of heat, then, according to the " principle of utilization," the steam-engine ; first Hertz's playing around
with electromagnetic waves, then radio and radar.   The
Second World War was perhaps the first time when the
need arose for the atomic bomb—" applied research "—
and this led to a rapid advance of atomic physics, and thus
of " fundamental research " ; a development the benefits
of which for mankind are highly questionable.   The progress in science and technology which signalizes our
civilization—but only this civilization of ours, not the
ancient, the Indian, or Chinese—is the expression of an
inherent and probably tragical dynamics.   It rolls ever
on, carrying us with it, and it progresses not because it
profits the individuals, the nations, or mankind, but
" according to the law that hath it started " (Goethe).

So, evolution appears to be more than the mere product
of chance governed by profit.   It seems a cornucopia of
évolution creatrice, a drama full of suspense, of dynamics
and tragic complications.   Life spirals laboriously upwards to higher and ever higher levels, paying for every
step.   It develops from the unicellular to the multicellular, and puts death into the world at the same time.
It passes into levels of higher differentiation and centralization, and pays for this by the loss of regulability after
disturbances.   It invents a highly developed nervous
system and therewith pain.   It adds to the primeval
parts of this nervous system a brain which allows consciousness that by means of a world of symbols grants
foresight and control of the future ;   at the same time
it is compelled to add anxiety about the future unknown
to brutes ;   finally, it will perhaps have to pay for this
development with self-destruction.   The meaning of this

play is unknown, unless it is what the mystics have called God's attaining to awareness of Himself.

From the standpoint of science, however, the history of life does not appear to be the result of an accumulation of changes at random but subject to laws. This does not imply mysterious controlling factors that in an anthropomorphic way strive towards progressive adaptation, fitness, or perfection. Rather there are principles of which we already know something at present, and of which we can hope to learn more in the future. Nature is a creative artist; but art is not accident or arbitrariness but the fulfilment of great laws.

6. *The Historical Character of Life*

Organisms are characterized by three principal attributes: organization, dynamic flow of processes, and history. As has been stated before, " life " is not a force or energy that, like electricity, gravity, heat, etc., is inherent in, or can be imparted to, any natural body. Rather it is limited to systems with a specific organization. Equally characteristic is the continuous flow and the pattern of processes in the organism. And finally, every organism originates from others of the same kind, and carries traits of the past, not only of its own individual existence but also of the history of the generations which preceded it. We shall later try to define the living organism in terms of its fundamental characteristics as a " hierarchical order of systems in a steady state." This definition, however, omits an important characteristic about which we can say little in an exact way, but which must not be completely lost from sight.

In physical systems events are, in general, determined by the momentary conditions only. For example, for a falling body it does not matter how it has arrived at its momentary position, for a chemical reaction it does not matter in what way the reacting compounds were produced. The past is, so to speak, effaced in physical systems. In contrast to this, organisms appear to be historical beings. For instance, when the human embryo displays gill clefts at a certain stage, it reveals that mammals evolved in geological times from fish-like creatures. In a similar way we find "historicity" in organic behaviour; the reaction of an animal or a human depends on the stimuli and reactions that the organism met with or produced in the past. This characteristic led Hering to the assumption of "Memory as a General Function of Organic Matter," to the mnemonic theories of life of Semon, Bleuler, and Rignano, to the parallelization of evolution and memory in individual organisms.

It is true that dependence on the past is not completely absent in certain physical phenomena. There are the phenomena of hysteresis, as found, for example, in remanent magnetism, elasticity, and the behaviour of colloids. Thus colloids liquefied by heat may solidify again; but if they are repeatedly liquefied the melting point is lowered; their behaviour is thus dependent on the previous history. Rashevsky[1] showed that the phenomena of hysteresis are based on the fact that the systems in question have several states of equilibrium, that is, conditions of minimum free energy. In this case the environment does not determine unequivocally the state of the system, for it can exist in different states of equilibrium in the same environment; which state is actually attained is determined by the previous history. Rashevsky has analysed these cases, so far little considered in theoretical physics on account of their limited bearing in inorganic events, from the thermodynamical and kinetic viewpoints, and investigated the "learning" in physical systems. According to him, systems with several equilibria have a number of properties that are characteristic of conditioned reflexes. Systems of this kind can respond conditionally to spatial and temporal "transformation patterns" of the environment. From these considerations Rashevsky and his co-workers have arrived at a far-reaching theory of brain mechanisms and of behaviour.

[1] N. Rashevsky, *Mathematical Biophysics*, Chicago, 1938; 2nd ed., 1948.

Mathematical expressions for dependence on the past are well known. Such cases are to be treated by means of integro-differential equations; in the equation defining the change of the system, an integral with respect to time appears which expresses the changes that have occurred during the previous history of the system. The principles of this " after-effect physics " were developed by Volterra and Donnan.[1]

However, the fundamental historical characteristic of the organisms is not included in such theory; namely, that in the course of phylogenesis, *anlagen* have accumulated which unfold progressively during ontogenesis according to Haeckel's biogenetic law, which, though to be modified in detail, is correct in principle. In the human embryo there lies the phylogenetic past from the protozoon through fish, amphibian, reptile, and primitive mammal, comprising hundreds of millions of years, but now repeated in nine months. This double process of phylogenetic accumulation of *anlagen* and their ontogenetic unfolding is rather to be compared with a gramophone disc on which traces or " engrams," fixed during recording and corresponding to the melody presented, are re-transformed into sound when the record is played. However, genetics makes no statements about the nature of the gramophone disc, called " genome."

Genetics and the experimental evolution theory are concerned exclusively with mutative changes in already existing genes. But obviously evolution comprises not only changes in existing genes but also creation of new genes. Otherwise we would come to an absurd preformism, which had to assume that the genes present in man have existed already in the primeval amœba. We know practically nothing about the origin of new genes beside occasional duplications, such as in the *Bar* mutation of Drosophila (p. 82), from which far-reaching conclusions can hardly be drawn. Likewise we can say

[1] V. Volterra, *Leçons sur la théorie de la lutte pour la vie*. Paris, 1931.—F. G. Donnan: " Integral analysis and the phenomenon of life." *Acta biotheoretica* 2/3, 1936/37.

hardly anything about the absolute number of genes in various organisms. The chromosome numbers have no definite relation to the taxonomic or phylogenetic position of the species. Considerations of this sort hold, at least in so far as the genes are interpreted as separate units. In a unitary conception of the basis of heredity, it may be possible to interpret phylogenetic changes not as an adding of new genes but rather as a transition to a new state in the genome as a whole; in a similar way that psychological memory is to be interpreted rather as a change in the whole " brain field " than as the accumulation of separate and individual traces (pp. 191 f.).

A further question comes in here. According to the second law of thermodynamics, the general direction of (macro-) physical events is towards a decrease of order and organization (cf. p. 178). In contrast to this a direction towards increasing order seems to be present in evolution. The author has referred to this important characteristic long ago (e.g., 1932, p. 64); Woltereck has called it " anamorphosis." In physical processes, chance and statistical probability act towards the levelling down of differences, as in the establishment of thermic equilibrium and in the dissipation of energy according to the law of entropy, due to the random movements of the molecules. In the biological sphere, on the contrary, according to the theory of selection, chance would act in the direction towards increasing differentiation and complexity.

Possibly three points are to be considered here. First, the law of entropy does not exclude the transition towards higher order. For example, crystallization which creates a level of organization higher than that of the molecules, of course follows the law of entropy; it is possible because of the existence of spatial vectors—the valencies or lattice forces present—and the law of entropy states only that the free energy in the system as a whole, namely, crystal plus solution, must decrease. If in organic systems organizational forces, " forces of crystallization of a higher order," are present, then their anamorphosis would not be in conflict with the law of entropy. Second, we may think of a profound difference between macrophysical and microphysical events. According to the second law of thermodynamics, macrophysical events are directed towards the destruction of existing order; but in inner-atomic and cosmic events, processes leading to higher order can take place according to the laws of quantum physics. For instance, in the interior of stars the formation of higher elements instead of radioactive decay, and thus an "anamorphosis" is possible. Perhaps, also, biological anamorphosis is ultimately to be considered from the viewpoint of quantum physics, as it is probably true for mutations (pp. 95, 165 f.). Third, there is a point discovered only recently and of fundamental importance. In contrast to the general trend in closed systems, a decrease of entropy can take place in open systems, and with it a transition to states of higher heterogeneity and complexity (p. 127).

An organism represents a spatial whole manifesting itself in the interactions of its parts and partial processes. In a similar way that the processes in the organism are determined by the whole spatial system (and not by isolated causal chains), it appears that they are also determined by the whole temporal context (and not only the momentary conditions). In the last resort, spatial wholeness and historicity are possibly different aspects of the same spatio-temporal whole. Within a four-dimensional way of consideration—where the time-dimension appears as a fourth co-ordinate besides those of space, and the world is frozen, so to speak, into a four-dimensional "time landscape"—they would signify different aspects of the same reality. Neither in space nor in time would the behaviour of a living system be determined by one-way causality (in the sense that the organism represents a sum of isolated causal trains, which are determined by momentary conditions), but rather by the whole space-time pattern. Such conception has certain analogies with the statements of wave-mechanics. If we were ever able to put the whole process in an organism into one formula, it would be an integro-differential equation, indicating the spatial and temporal whole at the same time. These are profound problems, which would have to be dealt with in connection with theoretical physics and General System Theory (pp. 199 ff.).

All these questions can only be stated; at present we have neither the factual foundations nor the theoretical means to give precise answers.

7. *The Nervous System : Automaton or Dynamic Interaction*

When we are undergoing a medical examination, one of the first things the doctor does is to tap the knee-tendon of the loosely hanging leg with a small hammer; the leg kicks upwards. What the physician is probing is a certain reflex arc. In the knee-tendon there are receptors which are stimulated by the tap. The excitation flows through a nerve path, the sensory nerve, into the spinal cord. Here are centres which switch the excitation from the sensory nerve to a motor nerve. Through the latter, excitation flows to a muscle, which it causes to contract, and the leg kicks up in the knee jerk or knee-tendon reflex. If the reaction is lacking, it means that there is a disturbance in the reflex arc. Hence the test of this and other reflexes is basic in the diagnosis of disorders of the nervous system the symptom of which is the loss of certain reflexes.

The field of the processes in the central nervous system and of behaviour, equally important from the standpoints of biology and medicine, belongs to those whose recent developments demonstrate impressively the rise of organismic conceptions.

The classical theory of centres and reflexes, as founded by the great neurologists of the nineteenth century, was an attempt to resolve the nervous system into a sum of apparatuses for definite functions and similarly animal behaviour into separable processes occurring in those structures. In this way the spinal cord was considered as a column of reflex apparatuses piled up in a segmental order. Again, the medulla oblongata contains many reflex centres and, equally, life-important automatic centres which need no external stimulus for their function, such as the respiratory centre which stimulates the breathing movements, the sugar centre, the centre which controls heart-beat, the vasomotor centres, and so on. Finally also the brain appears as a sum of central fields : motor areas of the cortex that correspond to the individual portions of body musculature and control their voluntary movements ; sensory areas which represent the apparatus of conscious perception of the different senses ; and association areas, in which the higher mental functions, especially memory and learning, are localized.

The theory of reflexes, centres, and localization in the nervous system is based upon an enormous number of facts, experimental as well as clinical. However, there are other well-known facts which indicate a great measure of regulability in the nervous system, and so are in contrast to that theory. For instance, paralysis caused by degeneration of the facial nerve can be cured clinically by grafting into the facial musculature fibres of the nerve serving the shoulder or the tongue (*N. accessorius* or *hypoglossus*). After some time the patient is able to control his facial musculature again, though the nerve supply is atypical. Or in a prosthesis after Sauerbruch, the artificial limb attached to the stump may be stretched by muscles normally functioning as flexors, and inflected by the extensors. Experience of this kind shows that nerves and centres are not fixed irrevocably and machine-like to a function. The same is shown by many experiments by Bethe, von Buddenbrock, and others. For instance, for the locomotion of insects, spiders, and crabs, the so-called cross-amble is characteristic. That is, in one phase the first leg of the left side, the second of the right, and the third on the left are put forward simultaneously, the opposite taking place in the following phase. This cross-amble is re-established, instantly and without a period of learning, if any number of legs is amputated. Then the co-ordination of the movements of the remaining legs is, of course, different from the normal. Therefore, it cannot depend on a fixed control mechanism, but must depend on the conditions present in the periphery and the central nervous system as a whole.

Thus we find an antagonism between the facts upon which the classical centre theory is based and those concerning regulation in the central nervous system. The latter are also of a high clinical importance, since they indicate the basis of possible restoration after disturbances. The author would like to reproduce here some of his earlier statements (von Bertalanffy, 1936, 1937), because these arguments, originally developed on the basis of the organismic conception, have been fully confirmed by later research. Thus they give a good demonstration of the value of this conception as a working hypothesis.

" The classical reflex and centre theory considered the central nervous system as a sum of isolate individual mechanisms. This conception could not have arisen, however, if the theory had not started from experiences with adult humans. Here we actually find, to a wide extent, fixation of the regions of the central nervous system to definite functions. Destruction of the lumbar cord abolishes the knee jerk; destruction of the visual centre in the brain leads to cortical blindness; a prick in the respiratory centre stops the breathing movements and causes death, and so on. On the other hand, the clinical and experimental facts of regulation show that this fixation is not absolute, and that the nervous system cannot be considered simply as a sum of fixed reflex apparatuses. This antithesis, however, becomes intelligible when we look at the phylogenetic and ontogenetic development of the central nervous system. Then we find that the principle of progressive mechanization applies. The central nervous system progresses from a less mechanized to an increasingly mechanized state, where it behaves to a wide extent like a sum of fixed mechanisms, though this mechanization is never complete, as demonstrated by the phenomena of regulation. A progressive fixation of the centres in phylogeny can be observed in the vertebrate series. In the lowest vertebrates, only few and vaguely defined centres can be found (*Myxine*, Herrick, 1929). In monkeys the motor areas are much less sharply defined than in apes. In the latter isolated movements of the fingers can be caused by electrical stimulation of certain points of the cortex; this is not the case in monkeys (Sherrington, 1907). Similarly, the ontogenetic development of the central nervous system shows that it is not local reflexes that are primary, as was supposed in the classical scheme; rather these crystallize out from primitive movements of the body as a whole or of larger body regions. This was demonstrated in very different species, such as embryos of the axolotl, the cat, bird, and man (e.g., Coghill, 1930; Coronios, 1933; Herrick, 1929; Kuo, 1932).

" Hence the following picture of the development of the nervous system can be outlined. Its original state is such that it behaves to a wide extent as an equipotential system. Then certain parts gradually take over more and more specialized functions, in a similar way as in development the embryo is at first equipotential, and gradually organ-forming regions are determined that are fixed to definite functions and can produce only one single organ. Similarly, definite reflex arcs segregate from original movements of the body as a whole. However, the capability for performing other functions is not completely lost. Even the classical theory was forced to hypostasize ' subsidiary centres,' which come into action when the main centre is no longer operative, such as respiratory and vascular centres in the spinal cord. The centres are thus not sharply circumscribed regions, but their function can be performed to a certain extent by larger regions of the central nervous system. Under normal conditions, however, each function is controlled by that part which ' knows it best.' If these ' controlling parts ' are destroyed, then other parts may take over the function, though with less efficiency. This principle is seen most convincingly in the function of the heart.

I

Normally, the sinus node (a system in the auricle acting as pace-maker of the heart-beat) is the controlling part and regulates the heart-beat. When it is prevented from doing so by complete heart block, then the atrioventricular node (on the border between auricle and ventricle) takes over the control; heart-beat goes on, though at a slower rate in the so-called Tawara-rhythm. Finally, even the His' bundle (the last branches of the pace-maker system of the heart) can also cause heart-beats. Similar principles obviously hold for the nervous system. The co-ordination between sense organs, nerve centres, and effector organs leads to the establishment of definite reflex paths that normally respond stereotypically and in a machine-like way to corresponding stimuli. But this specialization is not absolute, and the nervous system shows regulability because of its original equipotentiality, which was restricted but not completely lost during development.

" Clinical and experimental evidence concerning regulation in the nervous system shows that the function of its individual parts depends on the actual state of the system as a whole and on the relations with peripheral organs (sense organs, muscles). Thus, in beetles or crabs that have lost one or several legs the co-ordination within the nervous system depends on the number and arrangement of the legs still present. If organs are supplied by atypical nerves, the function of the paths and centres may change. But also the reaction of an intact animal is determined not only by the action of a definite centre alone but also to a greater or lesser extent by the state of the nervous system as a whole. Because of this, many reflexes can be shown clearly only in the decerebrated animal, not in the intact one, where they are profoundly modified by the interaction within the system as a whole. If, however, the function of the parts depends on the state of the whole system, it follows that the co-ordination in the reaction is not, or not completely, determined by fixed arrangements, but is controlled by dynamic laws within the system as a whole.

" The reflex theory considered the reflex arc, that is, the response to external stimuli, as the primary element of behaviour. On the contrary, recent research shows that it is rather autonomous function which is to be considered as primary. Automatic organs, such as the heart, and centres, such as the respiratory centre, are examples. The reactive mechanisms (reflex arcs) appear to develop, quite generally, on the basis of primitive rhythmic-locomotor mechanisms. This is shown, for example, in the fact that the sensory neurone takes part in the control of movements only after the motor neuron is already functioning. Thus movements appear in axolotl larvæ before motor and sensory nerve cells are connected; hence these movements cannot be reflexes to external stimuli, because the receptor part of the reflex arc is not yet connected with the motor part. They are automatic actions produced by the motor cells themselves (Coghill, 1929, 1930). In the human embryo likewise the first movements are of an automatic nature (Langworthy, 1932).

"The conception outlined applies not only to the functioning of individual organs but also to the behaviour of the organism as a whole. In many organisms the normal state in a homogeneous environment (absence of external stimuli) is not rest but speedy, vivacious movement. Similarly, we find this autonomous activity in instinctive behaviour. It is manifest here as a ' drive ' to perform certain movements, a drive that appears in a certain physiological state without external stimulus and causes, say, the quest for food or for a sex partner.

" These insights lead to revision of the notion of ' stimulus.' If the organism is primarily an active system, we must say : the stimulus (that is, an alteration of external conditions) does not cause a process in an internally inactive system, but rather modifies the process in an internally active system. This leads to the important conclusion that in the end it is not so much the external influences, the stimuli, but rather the internal situation, the distance from a normal state—the ' need,' psychologically speaking—that determines the reactions of the organism. This corresponds to the actual state of affairs. The organism is motivated in the first place not by stimuli but by needs to look for food, for a mate, and so forth. These 'drive actions' last until they lead to a situation that restores the normal state. There is only a difference in degree between drive movements caused by internal conditions without external stimuli and reactions caused by external stimuli; also the latter depend largely on the physiological state, as when at sight of prey a hungry animal responds with attack, but if satisfied pays no heed to it. Pflüger (1877) made the classic statement : ' The reason for the need is the reason for the satisfaction of the need.' In this he intended to express the soul-like purposiveness peculiar to living things. However, nothing vitalistic or psychological is expressed in the statement. It merely says that removal from a state of physiological equilibrium leads to actions that last until the normal state has been restored.

"We also find in instinctive behaviour a progressive mechanization or determination, quite similar to that in embryonic development. An excellent example is seen in the larvæ of the caddis-flies, which build pretty cases out of stones, wood, pine-needles, and the like (Uhlmann, 1932). Different types of these cases can be distinguished : a primary one, in which the tube inhabited by the animal is just loosely assembled; a secondary, with a coherent but irregular arrangement of the building materials; and a tertiary, where the larva works on the building materials and arranges them brick-fashion into very neatly manufactured cases. In primitive species, there is 'pluripotentiality,' that is, they can use different materials; in highly specialized species, there is ' unipotentiality,' that is, only one special building material can be used. In the same way, in individual development there is first multipotentiality, which becomes gradually restricted in older larvæ in progressive specialization to a special kind of architecture. If the case is lost a new one is made, and this can be induced up to sixty times with the same animal. These renewed structures go through the same stages of development as the normal one; individual experience of the animal plays no part. We thus see that the essential traits of instinctive behaviour correspond astonishingly closely with the laws of embryonic development."

Thus, the field of stimulus-response phenomena and of behaviour shows especially clearly the necessity of the previously mentioned (pp. 18 f.) organismic points of view. In contrast to an analytical and summative conception, we have the dependence of processes on the whole system. In contrast to structural and machine-like order, we find the primacy of dynamic order and the principle of progressive mechanization. In contrast to the conception of the primary reactivity of the organism, its character as an active system is primary.

Recent research has fully confirmed these views. On the basis of careful and extended experiments von Holst (e.g., 1937) arrived at a " new conception " of the activity of the nervous system. According to this author, locomotor activities, such as the crawling movements of worms, the running movements of Articulata, the fin-movements of fish, etc., are caused by central automatisms that do not need external stimuli. Such movements can therefore also be controlled by the " disafferentiated " nervous system, that is, after the connections with the sensory nerves have been severed. In contradistinction to the theory of fixed reflexes, it is further shown that the co-ordination of movements is not rigid but plastic, and is controlled by dynamical principles, by interactions within the nervous system. To these principles belong " relative co-ordination " and the " magnet effect," that is, the tendency exerted by one automatism, the movement of a fin for example, to impose its rhythm or a definite phase-relationship upon another. Thus, according to von Holst, the reflex is not the primary element of behaviour but a device for adapting the primary automatism to changing peripheral conditions. K. Lorenz, in his work on instinctive behaviour, arrived at a corresponding conception of the primary nature of autonomous activity. In instinctive action, pre-established successions of impulses, the so-called hereditary co-ordinations, often manifested without a releasing stimulus and only modified by external stimuli, play a dominant role. The primarily automatic character of instinctive actions is shown especially in "running-idle reactions" occurring in the absence of stimuli under certain physiological conditions, for example, when a bird which has no material to build with performs the movements of nest-building in the open air.

Conclusions similar to those to be drawn from the physiological investigation of the reactions and behaviour of animals are reached in the psychological investigation of perception and mental life. The latter development is centred around *gestalt* theory (pp. 189 ff.). Von Holst also emphasizes the far-reaching conformity which exists between the dynamical principles of nervous co-ordination and experienced *gestalten*.

The conformity among the principles of regulation in embryonic development, those of the distribution of excitations in the nervous system, and of *gestalt* perception is equally interesting. For instance, destruction on both sides of the visual centres of the brain leads, as mentioned, to cortical blindness. Destruction on only one side produces hemianopsia, loss of half the field of vision in both eyes. This can be readily diagnosticated in the laboratory by suitable methods, the perimetric measurement. In ordinary life, however, the patients do not experience only half a field of vision but a whole one, though it is smaller than normal; this regulation is possible by the formation of a new spot of most distinct vision which is displaced in comparison to the normal. Thus, the remaining part of the cortex can do almost the same job as the intact apparatus previously did; in the same way as in Driesch's experiment the half system can perform the same action as the whole (Goldstein). The correspondence between original equipotentiality and progressive determination in embryonic development, and in the function of the nervous system, has already been mentioned. Similarly, the principle of " controlling parts " can be seen in both fields: in embryonic development in the form of organizers; in the nervous system in the centres.

CHAPTER FOUR

LAWS OF LIFE

The hand of yours that once so nimbly
Moved to do a deed of grace—
The structured form is there no longer,
Another now is in its place.
All is changed. The new hand, bearing
Now the name the other bore,
Came like a wave that rose and, falling,
Joins the elements once more.

GOETHE, Duration in Change.

1. *The Stream of Life*

" You cannot step into the same river twice; for fresh water is for ever flowing towards you." Out of the breaking dawn of antiquity rang this saying of Heraclitus of Ephesus, whom his contemporaries called " the obscure." It is easy to understand that Heraclitus should seem like a foreigner to the Greeks. The Greek world was dominated by the Apollonian ideal of static repose, as expressed in their marble sculptures as well as in Plato's ideas as plastic archetypes of the things in existence. Heraclitus, however, is the Dionysian thinker who regarded a ceaseless stream of events as the essence of reality. But that which alienated him from his own world brings him near towards ours. The gods of the North are not classic marble statues in the clear sunlight; they reveal themselves in the thunderstorm. And these fundamental tendencies expressed in ancient myth remain active in the abstractness of later science. Thus to the Greeks the atoms were small hard bodies seen with a sculptor's eye; physics of the western world has resolved them into a play of forces, nodal points of a wave dynamics.

The river, everchanging and yet everlasting, seemed to Heraclitus to typify the world. Not only the world around us—this is what Heraclitus wanted to express—but even we ourselves are not the same from one moment to the next. With this Heraclitean thought we put our finger on a profound characteristic of the living world.

When we compare inanimate and animate objects we find a striking contrast. A crystal, for example, is built up of unchanging components; it persists with them perhaps through millions of years. A living organism, however, only appears to be persistent and invariable; in truth it is the manifestation of a perpetual flow. As a result of its metabolism, which is characteristic of every living organism, its components are not the same from one moment to the next. Living forms are not *in being*, they are *happening*; they are the expression of a perpetual stream of matter and energy which passes the organism and at the same time constitutes it. We believe we remain the same being; in truth hardly anything is left of the material components of our body in a few years; new chemical compounds, new cells and tissues have replaced the present ones.

We find this continuous change at all levels of biological organization. In the cell a continuous destruction of the chemical compounds composing it is taking place, in which destruction it persists as a whole. In a multi-cellular organism cells are continuously dying, and are replaced by new ones; but it persists as a whole. In a biocœnosis or a species, individuals die and new ones are born. Thus every organic system seems persistent and stationary when considered from a certain viewpoint. But what appears to be persistent at one level is in fact maintained in a continuous change, formation, growth, wearing out, and death of systems of the next lower level: of chemical components in the cell, of cells in a multicellular organism, and of individuals in a biocœnosis.

This dynamic conception of the organism can be counted among the most important principles of modern biology. It leads to the fundamental problems of life, and enables us to explore them.

From the standpoint of *physics* the characteristic state in which we find the living organism can be defined by stating that it is not a closed system with respect to its surroundings but an *open system* which continually gives up matter to the outer world and takes in matter from it, but which maintains itself in this continuous exchange in a *steady state*, or approaches such steady state in its variations in time.

So far physical chemistry has been concerned almost exclusively with processes in closed systems. Such processes lead to chemical equilibria. Chemical equilibria are also basic for certain processes within the organism. For example, the transport of oxygen from the lungs to the tissues is based on the chemical equilibrium between oxygen, hæmoglobin, and oxyhæmoglobin : in the lungs, where there is a high oxygen tension, blood is charged with oxygen, which combines with the hæmoglobin to form oxyhæmoglobin. In the tissues, at a lower oxygen tension, the oxyhæmoglobin dissociates and oxygen is released. Here a chemical equilibrium is reached because the processes concerned are of a high reaction rate. The organism as a whole is, however, never in true equilibrium, and the relatively slow processes of metabolism lead only to a steady state, maintained at a constant distance from true equilibrium by a continuous inflow and outflow, building up and breaking down of the component materials.[1]

Thus, we must ask for an *expansion and generalization of kinetics and thermodynamics*. The theory of reaction kinetics and equilibria in closed systems, as offered by physical chemistry, must be supplemented by a theory of

---

[1] For this state the author has introduced the term *Fliessgleichgewicht* (von Bertalanffy, 1942). The introduction of this notion is advisable because there is, in German, only the term " stationary," which is used in different ways : closed systems in which processes are going on, such as atoms or chemical equilibria, are termed " stationary," as well as open systems, such as a regulated jet of water or a flame. Therefore the author proposed distinguishing between *true equilibria* in closed systems and *fliessgleichgewichte* in open ones. In English, however, there is the well-defined term *steady state* for the latter.

open systems and steady states. After preliminary biological characterizations of the problem (e.g., von Bertalanffy, 1929, 1932, 1937), the present author has posed the physical problem, and has developed a number of principles of open systems and their significance for biological phenomena (1934, 1940, 1942). The notion and theory of steady states were adopted by German-speaking authors (e.g., Dehlinger and Wertz, 1942; Bavink, 1944; Skrabal, 1947); on the other hand, the problem was investigated by American and Belgian authors. An essentially similar treatment, with a number of further conclusions, was given by Burton (1939). The work of Reiner and Spiegelman (1945) appears to be stimulated by an exchange of ideas between the first of these and the present author. Prigogine and Wiame (1946) have supplemented the kinetic treatment of the problem with the thermodynamic.

The theory of open systems has opened an entirely *new field of physics*. " According to definition, the second law of thermodynamics applies only to closed systems; it does not define the steady state." This statement of the author (1940, 1942) involves far-reaching consequences, as the investigation of the thermodynamics of irreversible processes and open systems by Prigogine (1947) has shown. The author considers it as one of his most important achievements to have given from the biological side an impetus to this development in physics.

A full presentation of the author's treatment of the problem, as well as of the work of other investigators, will be given in Vol. II of this work. Prigogine states in his fundamental investigation :

" The two principles of classical thermodynamics hold only for *closed* systems, which exchange energy, but not matter, with the outside world, i.e., for very special systems. . . . Thermodynamics is an admirable but *fragmentary* theory, and this fragmentary character originates from the fact that it is applicable only to states of equilibrium in closed systems. *Therefore*,

*it is necessary to establish a broader theory, comprising
states of non-equilibrium as well as those of equilibrium.*"

We shall mention here only a few consequences of the
thermodynamics of open systems that open wide vistas
in physics as well as in biology, and partly overthrow
fundamental notions hitherto taken for granted. Whereas
in closed systems the trend of events is determined by
the increase of entropy, irreversible processes in open
systems cannot be characterized by entropy or another
thermodynamic potential; rather the steady state which
the system approaches is defined by the approach of
minimal entropy production. From this arises the revo-
lutionary consequence that in the transition to a steady
state within an open system there may be a decrease in
entropy and a spontaneous transition to a state of higher
heterogeneity and complexity. This fact is possibly of
fundamental significance for the increase in complexity
and order which is characteristic for organic development
and evolution (pp. 56, 64, 112 f.). Le Chatelier's principle
holds not only for closed but also for open systems. The
investigation of irreversible phenomena leads to the conception of a thermodynamical as opposed to astronomical
time (clock time); the first is non-metrical (i.e., not
definable by length measurements) but (under the simplest
assumptions) logarithmic; it is statistical because based
upon the second law; and it is local because it results
from irreversible processes at a certain point in space.

Just as the theory of open systems opens up a new
chapter in physics, it does the same in biology. Organisms have for a long time been called systems in " dynamic
equilibrium " to express the fact that they are maintained
in a state of perpetual exchange of their components.
However, only lip-service was paid to this problem, and
never a definition of the concept and of the principles
governing this characteristic state was given prior to the
work mentioned.

Living systems represent steady states of an extremely
complicated kind, consisting of an enormous number

of reaction components. The character of the organism, as being an open system, is at the basis of the phenomena in the living. To be sure, in considering the organism as a whole, we are not in a position to take account of all the individual reaction partners. However, as we shall see presently, it is possible to make statements on organic systems in the gross; and this leads to quantitative laws for important biological phenomena on the one hand, and to an explanation of basic attributes of life on the other.

Hierarchical organization on the one hand, and the characteristics of open systems on the other, are fundamental principles of living nature, and the advancement of theoretical biology will depend mainly upon the development of a theory of these two fundamentals.

" Whatever the nature of organizing relations may be, they form the central problem of biology, and biology will be fruitful in the future only if this is recognized. The hierarchy of relations from the molecular structure of carbon compounds to the equilibrium of species and ecological wholes will perhaps be the leading idea of the future " (Needham, 1932).

" The law of Guldberg and Waage governs chemical statics and dynamics. In organisms, we are confronted with chemical systems that have been for a long time designated as mobile or dynamic equilibria. Today the problem of the dynamic equilibrium of organisms no longer hangs completely in the air, but can be considered as the keystone and crown of a high and bold structure. The foundations for the construction are already laid by the physical chemists; further building is the business of physiologists just as is the testing of old tools and their improvement or the creation of new ones for special purposes, which will not always be served by a routine application of old methods " (Höber, 1926).

2. *The Definition of the Organism*

We may venture to give a tentative definition based on the considerations given : A living organism is a hierarchical order of open systems which maintains itself in the exchange of components by virtue of its system conditions.

This definition is certainly not exhaustive.   It neglects a third attribute essential to living systems, namely, their historical character (pp. 109 ff.).   With this reservation, however, it corresponds to the demands necessary in a scientific definition.

The number of attempts to define " life " is legion (cf. A. Meyer).[1]   In the first place we find pseudo-definitions, that is, definitions that introduce the definiendum in a concealed form into the definition, and thus imply a vicious circle.   For example, life has been defined as an " embodiment of forces that resist death," the notion of " death " being significant only as the opposite to " life."   Another group tries to define living organisms by enumerating their most important phenomenological characteristics.   An example is Roux's definition (1915), which lists the " auto-ergasies " or self-activities of the organism—such as autonomous transformation, secretion, restoration, growth, movement, division, inheritance, all of these functions being governed by self-regulation.   This may be quite correct as a descriptive characterization, distinguishing living from non-living systems ; it is not a definition in the strict sense.   Strict definition demands : (1) that it must not include characteristics of the definiendum ; (2) that it allows an unequivocal distinction from other phenomena ; and (3) that it provides the basis for a theory wherefrom special phenomena and their laws can be deduced.   Our present definition seems to satisfy these demands.

First, the definition does not include special characteristics of the organism which is to be defined ; but the basic phenomena of life can be considered as conse-

[1] A. Meyer, *Logik der Morphologie.*   Berlin, 1926.

quences following from the definition, as will be shown presently.

Secondly, the definition must state the conditions necessary and sufficient for a natural object to be called "living." The principles indicated are necessary; for an object without organization and without its maintenance in the change of components would not be a "living organism." On the other hand, they seem to be sufficient to distinguish between living and non-living systems. For example, a crystal shows hierarchical organization from elementary physical units to atoms, then to molecules and to crystal lattices; but it lacks maintenance through change of parts. In contrast, steady states in inanimate nature, such as a steady water-jet, a flame, a stationary electric current, show the necessary maintenance through change, but hierarchical organization is lacking, and their maintenance is guaranteed only by conditions outside the system, by a suitable "machine," such as a stopcock, a candle with wax and wick, and so on. Nevertheless, maintenance through change due to conditions within the system is not a vitalistic notion; it exists also in certain inanimate systems; this is shown by the drop systems investigated by Rashevsky,[1] which maintain themselves by intake and output of materials, grow, and divide. Here again, hierarchical organization is missing. If, however, in future but as yet unattainable experimentation, a system could be contrived which had all characteristics given in our definition—such as a colloidal system showing an inner organization, maintaining and multiplying itself by intake and output of materials—then we should not be able to tell whether it is a "living organism" or not. Viruses fall outside our definition because they are not capable of growth outside living cells, and hence do not possess all the conditions for autonomous maintenance through metabolism.

We shall also see that our definition fulfils the third

[1] N. Rashevsky, *Mathematical Biophysics.* Chicago, 1938; compare also the following volume.

condition, namely, that it can form the basis of a theory from which special laws can be deduced.

Is there a bridge or a gulf between the non-living and the living ? From macromolecular compounds and colloidal structures the way leads through elementary biological units with co-variant reduplication to the simplest cells, and thence to the manifold forms of plants and animals. We cannot point to an absolute break. Recent research rather seems to indicate that in viruses a more or less continuous inter-gradation exists from macromolecules, like the tobacco-mosaic virus, through molecular complexes, such as the big bacteriophages and the virus of polyhedra disease in insects, to very simple, organism-like forms, and finally to forms not far from bacteria, such as the waste-water organisms and rickettsias. On the other hand, passing from molecules, which are defined chemically, to the material and dynamic microcosm of the living cell, we come into a realm of entirely new phenomena. Although there is no absolute discontinuity, there is, if we can picture the transition in the form of a curve, a steep ascent from a lower to a higher level. The decisive point is not co-variant reduplication as such, but the attainment of a higher level of organization as well as the orderly pattern of a countless number of physico-chemical processes maintaining a steady state.

The basic physiological phenomena can be considered as consequences of the fact that an organism represents an open system. In general physiology the functions of life are arranged under three main headings. The first is physiology of *metabolism*, of the processes of catabolism and anabolism continually going on in the organism. The fundamental principle is the maintenance in a steady state. The second field is that of the phenomena of *irritability* and *motility*, including reactions to external stimuli as well as *autonomous activities* going on without external stimuli, for instance, heart-beat and breathing movements. These processes can be considered as waves superimposed on the steady state. A stimulus is a displacement from the steady state, into which the organism tends to return; the autonomous periodic activities represent smaller wave-trains superimposed on the continuous flow of the steady state. The third main chapter comprises the phenomena of *morphogenesis*, that is, the relatively slow changes in growth, development, senescence, and death. They are an expression of the fact that an organism is not truly stationary, as it can be considered with respect to the transitory processes of metabolism and irritability, but only quasi-stationary, that is, its steady state is slowly established and it undergoes slow changes.

3. The System Conception of the Organism—a Basis for Exact Biology

From the position now reached new insight can be obtained into many problems, leading to some extent to exact quantitative laws for basic phenomena of life. Only a brief survey will be given here of some of the conclusions to be treated more fully in the following volume.

In future *Bioenergetics* must be based on the theory of open systems. Any true equilibrium, and therefore any reaction system in chemical equilibrium, is incapable of doing work. For work to be done a gradient must exist; that is, the system must be removed from true equilibrium. Thus, a reservoir of water contains a large amount of potential energy, but this cannot drive a motor. For work to be performed, a gradient must exist, the water must flow downwards. And in order to maintain this performance we must arrange to have a stationary stream. The same applies to the organism. The chemical energy contained in its compounds cannot be used so long as they are in chemical equilibrium. As a system in a steady state, however, wherein reactions tending towards equilibrium go on continuously, the organism possesses the constant capacity for doing work which is necessary for the uninterrupted performance of its functions.

A further interesting problem arises here. If the organism is a system in a steady state, then a constant supply of energy is necessary for the maintenance of the distance from equilibrium. Therefore energy is necessary to the organism not only for the performance of its manifold functions, such as muscular and glandular activity and so on, but also for the maintenance of the steady state. This problem of the *maintenance work* of the cell and the organism as a whole is a basic question of bioenergetics, for which the theory of open systems provides the necessary principles (von Bertalanffy, 1942), which await experimental verification.

The fundamental problem of metabolism is its *self-regulation*. The difference between reactions in a living organism and those in a decaying corpse is that in the former all take place in such a way that their result is the maintenance of the system (pp. 13 f.). How does it happen that in a continuous stream of component materials the organism is maintained approximately constant, and in a state that does not correspond to an equilibrium based on reversible reactions, but in a steady state based on irreversible processes ? How is it that the materials worn out in the catabolic process are regenerated in the right way from compounds introduced as food, and that the compounds liberated from the latter by enzymatic action find the " right " place in the cell and in the organism, with the result that they maintain themselves in metabolism ? How is the constant composition guaranteed, even in spite of varying intake ? These main characteristics of self-regulation—constancy of a characteristic pattern of composition with continual change of building materials, independence of and persistence in changing supply, at different absolute sizes, restoration following catabolism, normal or enhanced by stimulation—are consequences of the general properties of open systems. Thus, the self-regulation of metabolism is, in principle, intelligible from physical laws (von Bertalanffy, 1940, 1942).

Possibly the application of the theory of open systems to " metabolizing crystals " gives an explanation of the problem of *co-variant reduplication* of elementary biological units (pp. 30 f.).

K

Before proceeding to further problems, some preliminary remarks have to be made. From the days of old, biology has been divided into two main fields, morphology and physiology. Morphology, the study of organic forms and structures, includes systematics as the distinction, description, and classification of animal and plant species, anatomy as the description of their structures, histology, cytology, embryology, and so on; physiology is the study of the processes in the living organisms, such as metabolism, behaviour, and morphogenesis. This subdivision is based upon the different methodology in the fields, technical as well as conceptual, and in this sense it is necessary. However, morphology and physiology are only different and complementary ways of studying the same integrated object.

The antithesis between *structure* and *function, morphology* and *physiology*, is based upon a static conception of the organism. In a machine there is a fixed arrangement that can be set in motion but can also be at rest. In a similar way the pre-established structure of, say, the heart is distinguished from its function, namely, rhythmical contraction. Actually, this separation between a pre-established structure and processes occurring in this structure does not apply to the living organism. For the organism is the expression of an everlasting, orderly process, though, on the other hand, this process is sustained by underlying structures and organized forms. What is described in morphology as organic forms and structures, is in reality a momentary cross-section through a spatio-temporal pattern.

What are called structures are slow processes of long duration, functions are quick processes of short duration. If we say that a function such as the contraction of a muscle is performed by a structure, it means that a quick and short process wave is superimposed on a long-lasting and slowly running wave.

A living organism is an object maintaining itself in an orderly flow of events wherein the higher systems appear persistent in the exchange of the subordinate

ones. This conception, first expressed, as far as we can
see, by the author, is well accepted. For example, it
was expressed by the anatomist Benninghoff (1935, 1936,
1938, 1939) in the following way, which corresponds
almost literally with the author's statements :

> " Thus, while within the body the components are
> in a state of flux, the body itself seems to persist.
> But also the individual represents a series of events
> that starts with fertilization and ends with death. . . .
> What is in a slow flow, and is relatively persistent and
> quasi-stationary, is impressive as an organic form,
> the quicker flow of events is the function maintaining
> that form. . . . If I look from the lower levels to the
> higher, then forms are apparent. The higher system
> acts as the form into which all subordinate events
> are integrated. Looking the other way, travelling
> down through the various levels, the forms are
> resolved one after another into processes whose speed
> increases with decreasing size of system." (1938)

In recent years, the conception of the organism as a
system in a steady state became familiar especially by
the advance of the tracer methods which have shown
that the processes of breaking down and synthesis of
building materials in the organism go on at a speed
hitherto unsuspected. Compare, for example, the statements given above with the following resumé of tracer
work :

> "The discovery and the description of the dynamic
> state of the living cells is the major contribution
> that the isotope technique has made to the field of
> biology and medicine. . . . The proteolytic and hy-
> drolytic enzymes are continuously active in breaking
> down the proteins, the carbohydrates and lipids at
> a very rapid rate. The erosion of the cell structure
> is continuously being compensated by a group of
> synthetic reactions which rebuild the degradated
> structure. The adult cell maintains itself in a steady
> state not because of the absence of degradative re-

actions but because the synthetic and degradative reactions are proceeding at equal rates. The net result appears to be an absence of reactions in the normal state ; the approach to equilibrium is a sign of death " (Rittenberg, 1948).

Thus, organic structures cannot be considered as static, but must be considered as dynamic. This first applies to the protoplasmic and cellular structures (pp. 32 ff.), and for the reasons given it is especially impressive at this level of small dimensions. Formations such as the nuclear spindle, the Golgi apparatus, and the like appear as structures when we have them before us in a fixed and stained microscopic preparation. However, if we consider them in their changes in time, they are a manifestation of processes at the chemical and colloidal levels, quasi-stationary states that last for a while but soon undergo changes or disappear.

In principle the same holds also for the macroscopic structure of the organism as a whole. Also in the latter, what is ultimately persisting is not a lasting structure but the law of a steady process.

This conception of the organism as the expression of a flow of events holds far-reaching consequences. It leads to a " *dynamic morphology* " (von Bertalanffy), the task of which is to derive organic forms from a play of forces controlled by quantitative laws. In this way the fields of metabolism, growth, and morphogenesis are integrated.

Growth is undoubtedly one of the main problems of biology. Indeed, growth has often been considered the central mystery of life. However, why an organism increases in size and why this increase slows down and finally ceases when the organism has " grown-up " are questions which physiology has so far failed to answer. But by regarding organisms as open systems, an exact *theory of organic growth* was developed, giving an explanation and quantitative laws for this basic biological phenomenon (von Bertalanffy, 1934–48, cf. Vol. II). In general, it may be said : an organism grows so long as anabolism outweighs catabolism; it becomes stationary when both processes hold the balance. Experience shows that catabolism of building materials is, in a first approximation, proportional to body-size. Anabolism of building materials, however, appears to depend, as far as higher animals are concerned, on energy metabolism, which yields the energies necessary for building up the organic components. In the most important case applying, e.g., to vertebrates, energy metabolism is surface-proportional. From these premises, laws of growth can be derived which make possible its calculation and the explanation of the peculiarities of the growth curves in the different classes of animals. In suitable cases these growth laws are valid with an accuracy comparable to that of physical laws. This has been demonstrated in various instances, ranging from bacteria and tissue cultures to fish, mammals, etc. In this way growth can be connected with, and derived from, total metabolism.

The theory has shed new light on many problems of which only the most important ones shall be enumerated here: the *absolute body-size* and the *explanation and calculation of growth in time of animals*; the *principle of the constancy of cell-size*; the *cyclic growth of mammals*; the *course of regenerative growth*; the verification of the theory by *measurement of the absorbing surfaces*; the *statement of different metabolic types of animals* with respect to the dependence of respiration on body-size and in correlation with corresponding *growth types* as deduced by the theory; the correlation between intensity of metabolism and body size with respect to *sex differences*; the *calculation of the intensity of catabolism of building materials* from the growth curves of animals, and verification of the calculated values by independent physiological experiments; the application of the theory to ecological problems, such as the dependence of *growth on temperature* (Bergmann's rule) and to *geographical variation*; the *peculiarities of the human growth curve*, and their significance for the somatic and mental development in man.

Another range of problems in dynamic morphology concerns the morphogenetic changes in development and evolution. If we compare organic forms it appears that their differences are based largely upon differences in proportions, which in turn depend on differences in growth rates, since some parts grow faster and others more slowly than the body as a whole. This applies, first, to individual development. For instance, the head of the newly born human is about a quarter of the length of the body ; but in the adult it is only an eighth of the body length. The head therefore grows more slowly in comparison to the body as a whole. The reverse holds for the legs : they grow faster than the rest of the body, for the newly born babe has relatively far shorter legs than the grown man. Thus body form is largely determined by the relative growth rates of the parts. In numerous cases this relative growth, the harmonization of growth rates, follows a simple quantitative law, namely, that of allometry stated by Huxley (1924). Similar considerations apply to evolution. Also the distinctions between related species depend to a great extent on differences of body proportions and thus on the harmonization of growth rates following the law of allometry.

In our own work a number of related problems was investigated, partly applying the principle of allometry to new fields. We may mention some of these problems : the *relation between body size and metabolism* and the statement of metabolic types based upon this relation ; the *dependence of rhythmic processes*, such as pulse and breathing frequencies, on body size, and its quantitative laws ; the relation of *metabolic gradients and growth gradients* ; a *critical examination of Child's theory of physiological gradients* ; a *synthetic theory* of the *absolute growth* of the body as a whole and *relative growth* of parts ; quantitative *laws of pharmaco-dynamical action* ; consideration of the significance of relative growth for the *types of somatic constitution in man*. Theoretically new outlooks are provided on fundamental questions of evolution, such as the directiveness or *orthogenesis* in evolution ;

co-adaptative changes in speciation; the *significance of hormonal factors* in evolution, and so on.

Yet another fundamental principle of morphology finds its physiological foundation in this connection. The classical morphology of the eighteenth century stated the " principle of the equilibrium of organs," which Goethe expressed as " budget law," and Geoffroy St. Hilaire as the " *loi de balancement*." It means that in the animal body there is a characteristic constant relationship between the sizes of the organs and, as we may add, also of their chemical components. The harmonization of growth rates according to the law of allometry is ultimately based upon a competition between the parts in the organism, every organ being able to assimilate a characteristic share of the nutritive materials gained by the organism as a whole. Hence it grows at a definite rate. The fact that allometric growth represents a process of distribution tending towards definite steady states presents the physiological basis of the *loi de balancement*.

Organic form appears to be the problem which is most recalcitrant to quantitative analysis. The modern approach shows, however, that it is controlled by quantitative laws which are gradually revealed. We may compare the dynamic conception in biology with that in physics. As in modern physics there is no matter in the sense of rigid and inert particles, but rather atoms are node-points of a wave dynamics, so in biology there is no rigid organic form as a bearer of the processes of life; rather there is a flow of processes, manifesting itself in apparently persistent forms.

Corresponding theories leading to quantitative laws are also to be found in the physiology of the senses and of excitation. Thus, the quantitative theory of sensation, especially vision, as developed by Pütter (1920) and Hecht (1931), is based upon principles similar to those expounded above. In the eye, as in a photographic plate, there exist light-sensitive substances which are decomposed by exposure. In the rods, which are sensitive to brightness, the substance in question is the visual purple. For photochemical reactions, the Bunsen-Roscoe law applies. It states the fact, well known to the photographer, that in order to obtain a certain effect, e.g., blackening of the photographic plate, the smaller the light intensity, the longer must be the time of exposure; or, in mathematical terms, the product of light intensity by time is constant. Unlike the photographic plate, however, the eye has a threshold below which light is ineffective. This is due to the fact that, counteracting the decomposition of visual purple into substances which eventually lead to nerve excitation, there is a second reaction which regenerates the visual purple from its decomposition products, and hence removes the excitation substances. From the principle indicated, the laws of light sensation, the existence of a threshold, adaptation to light and darkness, intensity discrimination, the Weber-Fechner law and its limitations, etc., can be derived quantitatively.

It is the characteristic energetic state of the organism which is also at the basis of that characteristic of excitation phenomena which can be called trigger action. Just as a great amount of energy is set free by a small spark in a keg of dynamite, the amount of energy set free in the response to stimuli, in the contraction of a muscle, for example, stands in no quantitative relation to the amount of energy, electric, mechanical, chemical, etc., of the stimulus. As we have said earlier (p. 119), it is not a machine originally at rest that is set going by a stimulus; rather the stimulus causes the discharge of energy stored when the organism was at rest. The organism could not perform its functions if, machine-fashion, it had to be set into action suddenly. It works on a much more economical principle, comparable with that of an accumulator, which during slack periods, say at night, accumulates a store of energy which can be set free on demand. Physiologically, this is shown by the fact that in nerve or muscle action a great part of the metabolic processes takes place not during the action phase but during recovery when the system is "re-charged"; it is just this which consumes energy. This leads to a unitary conception of both stimulus-response actions and the rhythmic-automatic actions needing no external stimulus (pp.119f.). Also in the latter, the basic principle is that of discharge and re-charge. The rhythmic-automatic actions follow the principle of relaxation oscillations or *kippschwingungen* (Bethe), which is applied, for instance, in advertising illumination; here a condenser system is gradually charged, and on reaching a critical potential, it discharges through a neon tube, followed by new charging, new discharge, and so on, so that the lamp flashes at rhythmical intervals. In a similar way, rhythmically active organs accumulate energy by metabolic processes and discharge it suddenly on reaching a certain level. Thus, the same principles are at work in automatic activities and in reactions to external stimuli. For this reason, there is no sharp borderline but rather all intermediates between typically reactive organs and nerve centres, which, like an isolated muscle or a reflex centre, enter into action only after external stimuli, and rhythmically active organs and centres, which, like the heart or the respiration centre, are active under constant external conditions. The basic phenomena of excitation, such as trigger action, the ratio of metabolic intensity in the initial and recovery phases, rhythmic automatism, and so on, are therefore consequences of the same principle, namely, that organic systems are not set going primarily by an outside influence, the stimulus, but are inherently active systems.

Although we are only at the beginning, we can say that the great realms of metabolism, morphogenesis, and irritability begin to merge into a unitary theoretical realm under the guiding principles of open systems and steady states. The impact of this is obvious.

If we look at physics, one of its most important achievements is the "homogenization" of reality, that is, the reduction of different phenomena to unified laws. It was doubtless a shock for the naturalists of the seventeenth century to have to realize that phenomena so qualitatively different as the orbits of the planets, the fall of stones, the swing of pendulums, tidal ebb and flow, should be governed by a single law, that of gravitation. Only later on is this shock overcome, and the unification of previously separate fields, such as of mechanics and heat in the kinetic theory, or of optics and electricity in the electromagnetic theory, comes to be considered the greatest triumph of science. A similar trend is apparent in modern biology. A vast complex of phenomena can be considered from the same point of view. In some branches it is already possible to express their laws in mathematical terms; in others we realize laws that at present can be indicated only in a qualitative manner, but follow the same conceptual scheme. Many fields can be subordinated to the same unitary conception, fields that are so diverse as certain physico-chemical phenomena which appear to be paradoxical within classical theory, but are explicable in the new conceptions, bio-energetics, metabolism, growth and dynamic morphology, the laws of evolution, the action of the sense organs and the nervous system, psychological *gestalt* perception, and so on.

Moreover, the new theory opens up the way to basic problems of the living world, in which the deepest secret of life was thought to be included.

Once more we return to the experiment which Driesch considered a proof of vitalism. The strange result of his sea-urchin experiment is indicated by the notion of equifinality. *Aequus*, the same, *finis*, the end : an equifinal event is one in which the same goal is reached from different starting points and in different ways. Apart from certain exceptional cases, we do not find equifinality in physical processes. Here a change in the initial conditions usually leads to a change in the final result : a damaged machine works differently from an undamaged one ; a change in the position of the barrel of a gun, in the quantity of powder used, changes the impact of the projectile, and so on. By contrast, equifinality is an important characteristic of processes in the living. In Driesch's experiment, for instance, the initial state may be different; for example, a whole germ, the half of a germ, the fusion product of two germs; nevertheless, the result is the same, namely, a normal larva. In the case of growth, equifinality can be formulated quantitatively. The same species-characteristic final size is reached from different initial sizes, such as different birth weights, or (when lasting damages are avoided) after temporary disturbance or stoppage of growth such as caused by a diet which is sufficient only for maintenance or is lacking in vitamins. Well, is equifinality a proof of vitalism? The answer is: No.

Analysis of the behaviour of open systems (von Berta-lanffy, 1940, 1942, 1950) shows that closed systems cannot behave equifinally; this is why, in general, we do not find such behaviour in inorganic systems. Open systems, on the contrary, which are in exchange of materials with the environment, and in so far as they attain a steady state, show the latter to be independent of the initial conditions, or in other words, they are equifinal. Equifinality is a necessary consequence of processes taking place in open systems in so far as they attain a steady state. Since there is in such systems a continuous inflow and outflow, building-up and breaking-down of the component materials, the steady state finally reached is not dependent on the initial conditions but only on the ratios between inflow and outflow, building-up and breaking-down. In other words, the final state does not depend on the initial conditions but on the system conditions which control the ratios just mentioned. For example, animal growth, according to the theory mentioned, can be explained as a result of the counteraction of catabolic and anabolic processes which are going on continually in the organism. Catabolism of building materials depends on the volume of the body; anabolism is, however, surface-dependent, at least in the most important type of animal growth. Now if a body increases in size without changing its shape, its surface-volume ratio is shifted in disfavour of surface, as may easily be visualized by comparing a roll to a loaf: the roll has much more crust, i.e., surface, in comparison to the soft part inside which fills the volume.

A similar thing is true for organic growth. As long as the organism is small, surface-dependent anabolism outruns catabolism, and so the organism grows. Eventually, however, a balance is reached wherein the synthesized materials only replace the materials that have been degraded in catabolism. Then the organism has entered into a steady state, is grown up. This steady state is not, however, dependent on the original size of the organism, but on the ratio between catabolism and anabolism of building materials characteristic for the particular species. Hence the same final size can be reached from different initial sizes or after disturbances. From this example, which can be treated quantitatively and submitted to calculation, we can draw an important conclusion : The directiveness which is so characteristic of life-processes that it was considered the very essence of life, explicable only in vitalistic terms, is a necessary result of the peculiar system-state of living organisms, namely, that they are open systems.

In this chapter we have hinted only in broad outlines and non-technical language at the vistas that are opened in the theory of open systems and its biological applications. In the following volume a more systematic presentation in mathematical language will be given. What has been said will suffice, however, to show that the theory of open systems and its application to living organisms can lead to new basic principles. This is true in two senses.

First, it allows the statement of exact laws for important life phenomena, such as metabolism, growth, morphogenesis, excitation, and sensory perception. Second, from the theory of open systems can be derived those general characteristics which seem most profoundly to distinguish the processes in the living organism from those in the inanimate world.

In a dramatic fashion those very phenomena are approached in the theory of open systems which, because of their apparent violation of physical laws, have been considered as "proofs of vitalism." Equifinality, Driesch's "first proof" of vitalism, appears as a result of processes in open systems. In a similar way, the self-regulation of metabolism—the maintenance and perpetual renewal of the cell by means of an interaction of innumerable reactions—has been considered as explicable only by assuming an entelechial factor (Kottje). This also becomes intelligible, in principle though not yet in detail, by application of the principles of open systems. According to the classical law of entropy, the natural trend of events is directed towards a chaotic state, characterized by maximum disorder, or, in other terms, towards thermodynamic equilibrium, where all processes come to a stop. In living organisms, however, we find a preservation of order and an avoidance of equilibrium. Hence from the standpoint of classical theory, as stated by Schrödinger, there exists only the possibility that an organism is a system that is governed not by the thermodynamic laws that result statistically from the principle of disorder, but by mechanical laws according to the principle of " order out of order." Schrödinger, however, clearly feels that this conception of the organism as a "mechanism" or "clockwork" is inadequate, as it is also refuted by organic regulation. Thus, there remains for him recourse only to an ego " that supervises the movements of atoms." Another formulation of the same argument is based upon anamorphosis : According to the law of entropy, the course of events is towards a degradation of order ; in the organic world, however, a transition towards higher order seems to take place ; and to explain this Woltereck appeals to "guiding impulses" of a "non-spatial Inner Life." In contrast, the thermodynamics of open systems inaugurates completely new points of view. Systems of this kind need not approach maximum entropy and disorder and a standstill of processes in thermodynamic equilibrium. Instead, spontaneous order, and even an increase in the degree of order, can appear in such systems. Yet another point is co-variant reduplication, the fact that genes and chromosomes divide but nevertheless " remain wholes." Actually, this was declared by Driesch to be the " second proof of vitalism." Probably this phenomena is also a consequence of the steady state of organic systems. Finally, Driesch's " third proof of vitalism " is based upon " action " and its " historical basis of reaction." It may be that this is also to be explained according to a system theory of memory (p. 191) connected with a dynamic conception of the function of the nervous system (pp. 118, 121).

So we shall hardly be mistaken in assuming that with these principles we are near the very roots of the fundamental biological problems.

CHAPTER FIVE

LIFE AND KNOWLEDGE

*The flow that even works and lives
Consolidate in thoughts perennial.*—GOETHE, Faust.

1. *The Whole and its Parts*

THE assertion that " the whole is more than the sum of its parts," that compared with its components it has " new " properties and modes of action, and the question whether higher levels of being can be " reduced " to lower ones are the essence of every " synholistic " theory or " unitary conception of the whole." Obviously here two statements are involved which are correct in themselves but antithetic.

On the one hand, every system in the hierarchical order, from the ultimate physical units to the atoms, molecules, cells, and organisms, exhibits new properties and modes of action that cannot be understood by a mere summation of the properties and modes of action of the subordinate systems. For example, when the metal sodium and the gas chlorine combine to form natrium chloride, the properties of the latter are different from those of the two component elements; similarly, the properties of a living cell are very different from the properties of the component proteins, and so on.

On the other hand, it is just the business of physics to explain higher levels in terms of the lower ones. Thus, the valencies exerted by an atom, and subsequently the compounds it yields in combination with other atoms, as well as other chemical properties, follow from the number of electrons available in the outer electron shell of the atom. Similarly, the spatial arrangement of the atoms within the molecule explains the configuration of the crystal formed by that compound. The structural formulæ in chemistry explain to a great extent those very properties which were considered as the prototype of the " non-summative," such as the characteristic colours of the hydrocarbons (compounds built up from elements which are colourless themselves), their taste, pharmacodynamical action, and so on. Hence the question arises as to what the supposed " non-summativity " of higher levels with respect to the lower ones really means, and in how far the former are explicable in terms of the latter.

The answer is simple. The properties and modes of action of higher levels are not explicable by the summation of the properties and modes of action of their components *taken in isolation*. If, however, we know the *ensemble* of the components and the *relations existing between them*, then the higher levels are derivable from the components.

Naturally, a mere summation of, say, a number of C, H, O, and N atoms gives no adequate knowledge of the compound molecule. This is readily seen, for example, in isomerism, when compounds consisting of the same atoms, but in different arrangement, have different properties. If, on the contrary, the structural formula is known, then the properties of the molecule are intelligible in terms of its parts, the component atoms. The same holds for every "whole." Adding up the charges in the parts of an electrical conductor, we cannot find the distribution of charge in the conductor as a whole, because it depends on the configuration of the whole system. If the parameters of the parts and the boundary conditions of the whole system are known, the distribution of charge in the latter can be derived " from the parts."

These statements are trivial. The truism that in order to know a given system we must know not only its " parts " but also the " relations " between them, that every system represents a " whole " or a *gestalt* (p. 192), became a problem and the starting-point of controversy only because of a misapplication of the mechanistic postulate in biology : the latter took into account the " parts " only, and neglected the " relations between the parts " (pp. 10 ff.).

However, there remains a problem which may best be illustrated by a few examples. The atom of an ideal monatomic gas can be considered, first, in mechanical heat theory as a material corpuscle subject to the laws of mechanics. Later on, it appears, in Rutherford's model, as a planetary system consisting of a central sun, the nucleus with its positive charge, and negative electrons as planets circling round it, the system being controlled by electrical forces, according to the law that the number of protons equals the number of electrons. The phenomena of radiation require further the introduction of quantum conditions, which are indicated by the model after Bohr. Finally, when we pass to the atomic nucleus, electrical forces no longer suffice. As a free particle, a proton has a positive charge. Nevertheless, the nucleus, consisting of protons and neutrons, holds together, although the protons in it should repel each other on account of their charges. Thus, if a proton is in a nucleus, nuclear forces are developed which are interpreted as exchange forces and must be taken into account in order to understand the nucleus of the atom. Another example : classical chemistry attributed to every atom a definite number of valencies, symbolized graphically like H⁻, ⁻O⁻, ⁻C⁻, etc., and saturated when the atom enters into chemical combination with another. In fact, these primary valencies are sufficient as far as chemical compounds in the classical sense are concerned. They are not sufficient, however, to explain, for example, crystallization, macromolecular compounds, cohesion, and so on ; rather does the atom display further forces termed secondary valencies, lattice or van der Waals forces. In turn, they are explained by modern electron and quantum theory. In all such cases the inclusion of new phenomena into physical theory necessitates a modification and refinement of the original picture, and it is just this which constitutes the advancement of physics.

Now the fundamental assumption of the so-called mechanistic conception in physics and biology was that all phenomena could be explained from a pre-established set

L

of laws. This was the ideal of the Laplacean spirit, according to which all events are to be reduced to "movements of atoms," i.e., to the laws of mechanics, which were considered as ultimate ; therefore the evolution of science would consist only in the application of these fundamental laws to all fields of phenomena. Actually, however, the advancement of physics tells a different and much more exciting story. Electrodynamics was never reduced to mechanics, nor quantum physics to classical physics. The inclusion of new fields of phenomena, and especially of organization, occurs in the way of a synthesis in which originally separate fields fuse into an integrated realm. This, however, is often carried through not by a mere application of principles stated in the beginning and simple derivation of higher levels from lower ones ; instead, the latter win new aspects when included into the generalized theory.

What has been said can be interpreted in a realistic or in an epistemological sense. In realistic interpretation one could say that in every system forces of a higher order are *potentiâ* present, which, however, become manifest only if that system becomes part of a configuration of a higher level ; if, for instance, the proton is part of an atomic nucleus, if the main-valency chain held together by " classical " valencies enters the micella of a polysaccharide, if a protein molecule becomes part of an elementary biological unit with self-duplication, and so on.

But this realistic or metaphysical interpretation mistakes the meaning of science. " Forces " are not metaphysical attributes inherent in certain physical configurations, but physics introduces " forces " as they are needed for the explanation and calculation of phenomena. Their meaning is that of visual models. What really matters, are formal relations, the system of the laws of nature. This system, however, tends to unification, that is, derivation of the special laws from the smallest possible set of primary hypotheses. To attain this goal, the primary assumptions must continually be altered and re-shaped in the course of the evolution of science.

This state of affairs we must bear in mind when considering the much-disputed problem of the relationship between physical and biological laws.

2. *Laws in Biology and Physics*

The discussion of the " mechanistic theory " in biology is much encumbered by the ambiguity of the term. The author has enumerated (1932) seven different meanings, and possibly this list is not exhaustive. The meaning of the term "mechanism" which is unequivocal is "non-vitalism," i.e., the exclusion of factors that are not accessible to scientific investigation and conceivable only by anthropomorphic empathy. In this sense, "mechanistic theory" is synonymous with natural science, and hence scientific biology must need be "mechanistic." As far as a narrower definition is concerned, there is, however, a great divergence of opinions. Some "mechanists," such as Bünning (1932), accept the recognition of specific biological laws as a matter of course. Others (Gross, 1930) consider the refutation of specific laws in biology as essential for the mechanistic theory, and in the same vein recognition of specific biological laws is called " vitalistic " by opponents to mechanism (Wenzl, 1937).

Obviously there are three different possibilities and questions to be distinguished here. These are : (1) whether biology is merely a field of application of the laws known in physics and chemistry; (2) if this is not so, whether biological laws can ultimately be reduced to, and derived from, physical laws ; (3) whether biological laws are of the same logical structure as physical laws.

It is manifest that biology, in so far as it is a *descriptive science*, is autonomous with respect to physics, and will always remain so, owing to the peculiarity of its subject-matter. Systematics, anatomy, morphology, embryology, biogeography, palæontology, physiological anatomy, ecology, phylogeny will not become branches of physics, even in a remote future. This is not because of the question whether biology is autonomous in its laws—which is no concern of these fields—but due to the simple reason that the number of forms and phenomena is incomparably greater in the living than in the non-living world. Descriptive mineralogy, for example, has become an appendix to physics and chemistry, because almost all is said by the indication of the chemical (mineral chemistry), morphological (crystallography), and physical (crystal physics) properties of minerals, and pure description, such as the listing of the different varieties of agate or feldspar, fades into the background. But the distinction between the vicious malaria mosquito *Anopheles* and the harmless gnat *Culex*, the circulatory systems in frog and man, the phylogeny of the saurians—such things can only be described. A future Newton of the grass blade, of whom Kant dreamed, will perhaps have a formula by which the wing-patterns of butterflies can be derived from a basic model by way of genetical and developmental analysis. But even so, he will not care to figure out the wings of the tens of thousands of butterfly species, with all their little dots and dashes, because he would have to appoint at least an equal number of assistants to do this job, which would, after all, be pointless. Even this zoologist of the future would acquiesce himself with definitions in vernacular language as they are found in the taxonomic literature of today. It must be emphasized that this unphysical procedure in biology is in no way limited to a humdrum description in the narrower sense of the word. Actually, the morphological comparison which leads to the establishment of a series of types, say of the vertebrate skull, the anatomico-physiological elaboration of the paths and reflexes of the spinal cord, the investigation of the phylogeny of man, and the inexhaustible abundance of similar questions, is based on specific biological notions, such as those of "type," "organ," "phylogenetic series," and the like; and they involve a system of order which we bring into the bewildering manifoldness of phenomena, just as it is done by the mathematical calculations of the theoretical physicist.

Just because we are championing exact, theoretical, and quantitative biology, we have to point out that what is expressed in the " exact " sciences as " laws " represents only a small section of reality. Even the greatest physicist will run after his hat when it is blown along in the street, and in doing so he is not concerned with the theory of heat, nor is he able to calculate the fickle eddy of the wind, convinced though he is that it obeys the kinetic theory. The geographer and meteorologist have no doubt that the formation of the earth's crust and atmospheric phenomena are based on physical laws, and that they are definitely not due to the labour of entelechial ghosts. Nevertheless, there are innumerable things in these fields that cannot be squeezed into a formula and can only be described, and where a rule of thumb must take the place of physical deduction. As mathematical biologists, we put the greatest emphasis on the obedience of organic forms to exact laws. We are extremely pleased to find that, say, the phylogenetic changes of the vertebrate skull follow the law of allometry (pp. 99 f., 138). But for precisely this reason we know only too well that it is only a small part of phenomena that can be understood in an " exact " way. Two skulls are distinguished not only by coarse differences of proportion, which we can measure and calculate, but also by a wealth of characters that can only be described in verbal language, or even are noticed only by the morphologist's trained eye, but which he is hardly able to put into words.

In this sense biology will never be " absorbed " into physics, and its place as an " autonomous science " *vis-à-vis* physics is obvious. This statement is outside the question of " biological mechanism " and quite independent of any decision about it. That question relates only to those general traits of order which we are able to state in the form of " *laws*."

Biology has the task of establishing the system-laws or laws of organization at all levels of the living world. These appear to transcend the laws of inanimate nature in two ways.

1. In the organic realm there exist *higher levels of order and organization*, compared to those in the inorganic.

Already in the configuration of macromolecular organic substances, and even more in the field of elementary biological units such as viruses and genes, we are presented with problems that go far beyond the structural laws of inorganic compounds.

2. The processes in the living are so complicated that with *laws concerning organic systems as a whole* we cannot take into account the individual physico-chemical reactions, but must use units and parameters of a biological order.   If, for example, the total metabolism of an animal is to be investigated, we cannot take into account the inconceivably numerous and complex reactions in intermediary metabolism; instead we must reckon with balance values, determining the overall result of all these reactions by way of oxygen consumption, carbon dioxide production, or calories production.   This is clinical routine in the determination of basal metabolism as a diagnostical method.   The same applies when we wish to establish quantitative laws for metabolism or growth; here also we have to use constants which are expressions for the lump result of innumerable physico-chemical processes.   In this way, it is possible to state global laws that are exact and deducible within a theory (p. 137). Similarly, genetics reckons not with physical processes but with biological units, such as genes, chromosomes in which the genes are located, plant or animal populations where the distribution of readily recognizable characters in successive generations is observed, etc.   In this way, genetics develops a system of statistical laws of admirable subtility and exactitude.   Again, the theory of population dynamics is one of the most advanced fields of quantitative biology, in its ecological (Volterra, D'Ancona, and others) as well as genetical aspects (Hardy, Wright, etc.).   Naturally, such theory cannot be stated in terms of physico-chemical units, but only in terms of biological individuals.   Laws of this kind are already established to a considerable extent in the several branches of biology, and it appears that their future development will render biology into an exact science.   Such laws are not " physi-

cal," for they refer to units that exist only in the biological sphere; but in fields that are sufficiently advanced they form a theoretical system the logical structure of which is the same as that of any field in physics.

The impact of quantitative laws is obvious. Understanding her laws is indeed the most important means of controlling nature. The development of modern technology and the control of inanimate nature have been possible only because exact laws were established, and future events can be calculated. In a similar way, the statement of biological laws will lead us more and more to a mastery of living nature.

It was often asserted that the statement of quantitative biological laws involves a reduction of biology to physics and chemistry. It seems hardly necessary to refute this. For mathematics is a tool capable of universal application, and can therefore be applied to any realm, the sociological or psychological for example, as well as to physics or chemistry.

There is a kind of complementarity between analytical and global treatment of biological systems. We can either pick out individual processes in the organism and analyse them in terms of physico-chemistry—then the whole, because of its enormous complexity, will escape us; or we can state global laws for the biological system as a whole—but then we have to forsake the physico-chemical determination of the individual processes.

The first procedure is the usual method of biochemistry, biophysics, and physiology. Experience shows, however, that the essential "vital" features seem to elude this attack. It is an ever-recurring refrain in biological literature that, in spite of extensive analysis of the physico-chemical factors concerned, the biological problem proper could not be grasped, and is waiting "for future investigation." For example, the investigation of the physico-chemical factors of permeability leads to the conclusion that they do not fully explain the import and export of substances in the living cell, and that in addition to "physical permeability" a regulated "physiological permeability " (Höber) or an " adenoid " activity of the cell (Collander) is to be assumed. This is obviously a semi-vitalistic notion which adds a regulatory factor to the physico-chemical processes. The correct interpretation is probably a system theory of permeability (von Bertalanffy, 1932) : The orderly and regulated transfer of substances as it is found in the living and metabolizing cell seems to be governed by the ensemble of factors present in the context of the organism as a whole.—The interpretation of protoplasm in terms of colloidal chemistry reaches its limit with the question why it is " living," that is, why it does not attain, like inanimate colloidal systems, a state of equilibrium, but maintains itself in a state of continuous change, of breaking-down, building-up, and regeneration.—Even the most detailed knowledge of the individual chemical reactions going on in the cell and in the organism does not answer just the fundamental question of organic metabolism, namely, its self-regulation, the co-ordination of the processes which guarantees maintenance in change of the components. This becomes intelligible in the theory of the organism as a system of reactions in a steady state (cf. the following volume).— Modern research has revealed the chemical nature of organizer action and gene-dependent substances. Automatically, however, the problem of development and heredity is shifted to the other side of the reaction complex, namely, to the organization of the substratum responding to those factors.—After an enormous number of investigations on experimental parthenogenesis, the real problem of the activation of the ovum remains unsolved, namely, the question what the various physico-chemical factors, besides physico-chemical changes in permeability, colloidal state, respiration, etc., actually do when releasing the amazing performance of the formation of a new organism. Considerations of this kind do not call in question the necessity of analytical investigation which is the basis for theoretical insight into the factors governing biological phenomena, as well as for fields of greatest practical impact, such as those of enzymes, hormones, and vitamins, of chemotherapy, and so forth. But they do show that as a complement the analytical procedure needs investigation of the organism as a whole and its global laws.

Thus, we shall answer the first question posed (p. 151) in the following way : biological laws are not a mere application of physico-chemical laws, but we have here a realm of specific laws. This does not mean a dualism in the sense that vitalistic forces enter the play in the living. But it appears that the level of biological laws is a higher one, as compared to that of the laws of physics (cf. pp. 172 ff.). A third level comprises the sphere of sociology.

Now the second question arises, namely, whether biological laws can ultimately be " reduced " to the laws of physics. The evolution of physics tends towards an ever more comprehensive unification which, though by no means finished, in principle lets us expect that the entire physical world can be constructed from a few ultimate elements and basic laws. From a small number of physical constants, such as Planck's quantum, the masses of the proton and electron, light velocity, etc., together with the pertinent fundamental laws, first the structure of the atoms and the periodic system of the elements can be derived, from these again the multiplicity of chemical compounds, the crystals, the rigid bodies, etc., up to planetary systems and galaxies. It can hardly be doubted that a fusion of the realms of physical and biological laws will ultimately be achieved. For from the standpoint of logics of science, the synthesis of previously separate fields is a general trend, and, on the other hand, from the empirical standpoint the realms of the sub-microscopic morphology, of viruses, etc., form a connecting link between inanimate and living nature. This basic postulate, however, does not obviate the necessity of first establishing the laws of the biological levels as such. On the other hand, it is possible, and even demonstrated to some extent, that the very inclusion of biological problems and fields leads to an expansion of the system of concepts and laws of physics. Remember the generalization of thermodynamics in the theory of open systems, which seems to contradict principles of the physical world hitherto regarded as fundamental, such as that of the tendency to maximum disorder. This is the more remarkable since thermodynamics seemed to be a consummate field of classical physics. Therefore, only the evolution of science itself will show in what way a synthesis can be achieved.

Number and measure govern mathematical physics, and pointer readings are its ultimate basis. In biology also the statement of quantitative laws is an important task, and we see that such laws can be found even in fields like that of organic forms. However, it appears that there is a range of specific biological problems, for the treatment of which the mathematical tools are yet to be created. Many of the most essential questions of biology are not a matter of quantities but of " pattern," " position," and " shape."

For instance, in the hierarchical order of the organism (pp. 37 ff.), it is not quantities that are interesting but relations of subordination and superordination, of centralization, and the like. In the morphogenetic movements (pp. 62 f.) it is not the numbers of cells nor quantities and mass relations that matter, but rather changes in relative positions; as, for example, when organ-forming regions or " fields " which are extended over the surface of the gastrula in the beginning contract in a certain way, win a definite position and shape in the embryo, and so on. Morphogenetic changes in certain spatial directions can be measured. In so far as they are based on relative growth, we find them governed by the law of simple allometry (p. 138). In this way the phylogenetic and ontogenetic change and increase of the equine skull, for instance, can be expressed in simple formulæ. Naturally, however, the changes of shape are not unidimensional, but follow numerous vectors in the different dimensions of space. This, again, can be expressed if we follow d'Arcy Thompson's method of transformation. If, say, the skull of the ancestral horse,

Eohippus, is projected into a rectangular Cartesian system of co-ordinates, by distortion of this co-ordinate system it can be transformed into the skull of the modern horse; in this transformation intermediate forms appear that correspond to phylogenetic stages in equine evolution. However, this is only a method of representation, and tells us nothing about the laws governing the transformation. What we would like to know is not equations for a few measurable vectors, but rather an integrative law that would show us why the transformation which has actually happened in the evolutionary series from *Eohippus* to *Equus* is singular with respect to the infinite number of other, mathematically possible transformations.

These are questions that, as far as can be seen, are partly related to topology and *analysis situs*; that is, they concern problems of relations within manifolds. They seem to be partly problems of group theory, since questions of invariance in the transformation of systems of equations appear. We may also think of developments in mathematical logic, as it was applied by Woodger for the definition of biological notions. Finally, General System Theory (pp. 199 ff.) will have a momentous share in future developments. These problems have in common that they are not of a quantitative nature but are concerned with relations of order and position.

It is usual to identify " mathematics " with the " science of quantities." This is correct so far as the general line of its evolution and its application to physics are concerned. In a wider sense, however, mathematics comprises every deductive system of order, and, as has just been pointed out, there are some beginnings of a " non-quantitative " mathematics. In this sense a few authors (von Bertalanffy, 1928, 1930; Woodger, 1929, 1930–31; Bavink, 1944), followed by others like Needham and Waddington, have pondered on the possibility that a non-quantitative or *gestalt* mathematics could have an important bearing for biological theory. It would be, as Bavink put it, a system of mathematics where not—as in ordinary quantitative mathematics, which fits so excellently the needs of physics—the notion of quantity but that of form or order would be fundamental.

It may be hinted at the example of physics, where new fields often had to develop their appropriate mathematics, new and unheard of till then, as was the case, for example, with the theory of matrices in wave mechanics.

> " When we remember that completely new mathematical developments were necessary for treating the most elementary systems in physics, developments that challenged mathematical physicists up to exhaustion, it would seem rather improbable that for the treatment of the most complicated systems in nature—the organisms—a mere application of routine physics and physical chemistry should be sufficient. Only by the close co-operation of biologists, theoretical physicists, mathematicians, and logicians will the mathematization of biology be achieved " (von Bertalanffy, 1932).

To be sure, this is " music of the future " and is intended only to indicate the tasks awaiting future generations of biologists. In any case, it is a fact documented by the history of science that progress is to a great extent dependent on the development of suitable theoretical abstractions and symbolisms. Only the evolution of analytical geometry and calculus made the development of classical physics possible. The relativity and quantum theories are linked with the development of non-Euclidian geometries, Fourier analysis, matrix calculus, and so on. The development of chemistry hinges upon the invention of the language of chemical formulæ. Similarly, genetics has become an exact field through Mendel's ingenious abstraction and the symbolism he created. On the other hand, the lack of a rigid theory in fields like developmental physiology is connected with the fact that they have not yet found the necessary abstractions and symbolisms.

Thus, the answer to the second question (p. 151) will probably be in the following line. As already explained (pp. 149 f.), the incorporation of new fields into physics is usually accomplished not by a mere extrapolation of given principles but in the way that first there is an autonomous development in the new field and that, in the final synthesis, the original field is also broadened. Chemistry was not developed by applying Newtonian mechanics to the atoms. First a new and specific world of constructs and laws was created; and unification was finally achieved because in the meantime the atom had metamorphosed from a mass-point to a complex organization. Biological "mechanism" presupposes a definitive catalogue of the physical laws of nature, which merely have to be applied correctly to the phenomena of life in order to explain them. But there is no such catalogue; and therefore we cannot foretell what extensions of the conceptual system of physics will be necessary before a final synthesis of both realms can take place.

The answer to the third question (p. 151) is unequivocal. The task of every science is to "explain." By explanation we understand the subordination of the particular to the general, and, conversely, the derivation of the special from the general. Hence the definitive form of science is a hypothetico-deductive system, that is, a theoretical construct in which by introduction of special conditions consequences that can be tested in experience can be derived from general statements. To a certain extent, this can be achieved by using vernacular language. However, the ambiguity of words, and the fact that their combination according to syntax does not always follow the rules of logical deduction, set a limit to the exactness of such a system. The goal of science can therefore be reached only when symbols having unequivocal and fixed meanings are linked according to equally unequivocal rules of the game. Such a system is called mathematics. In this sense Kant's statement that in every doctrine of nature there is only as much science proper as there is mathematics, is justified. For mathematics means nothing else than the highest attainable form of rationalization of reality.   For this very reason the mathematical formalism of modern physics, often reproved and leading to the non-visualizable character of its constructs, is neither arbitrary nor an evasion from a quandary but is a necessary concomitant of scientific evolution.   However, which form of mathematical theory will be suitable for the mirroring of reality in symbols we cannot tell *a priori*; only experience can decide.   In this respect modern physics is indeed not lacking in surprises.   Newton would probably have fainted had he been told that the fundamental laws of physics have not the form of differential equations with strict causal meaning but instead that of matrices and probability functions.   But whatever form the system of laws in future biology may have, and even if it included laws of a structure that we can guess only vaguely at present, it will bear the character of logical deduction, and hence of " mathematics," and will therefore be of the same formal character as physics.

3. *Microphysics and Biology*

A world governed by strict physical laws following the inexorable principle of causality ;   the ultimate goal of science to resolve all phenomena, including those of life and mind, into a blind play of atoms which bears no room for any purposiveness :   these were the foundations of that construct of the world which reached its apex in the nineteenth century and is called the " mechanistic theory." Its impressive symbol was the ideal of the Laplacean spirit, who, supposedly in possession of all laws of physics, would be able to calculate from the positions and velocities of the atoms at a certain moment the state of the universe at any other time, past or future.

It is one of the fundamental changes of the scientific world picture in recent times that physics has been disclosed as unable to state absolutely exact laws of nature but forced to acquiesce with statistical laws.

This knowledge was gained in two stages.   Classical physics has already discovered the statistical nature of the second principle of thermodynamics.   All directed energy is, compared to the unordered heat movement of the molecules, an improbable state. The transition of higher, directed forms of energy into undirected heat motion and the establishment of thermic equilibrium is therefore a transition to a more probable state, involving an increasingly uniform distribution of the molecules endowed with different kinetic energies. In this derivation of the second law as given by Boltzmann, it was not yet doubted that the path of every molecule is strictly determined by the laws of mechanics. In practice, however, because of the enormous number of molecules and their interplay, we must be content with a statistical law that states the average behaviour of a great number of molecules. This law is the second principle of thermodynamics, which states that, notwithstanding the multiplicity of molecular movements, the general trend is towards thermic equilibrium. In very small dimensions, however, departures from the probable distribution as required by the second law do occur. For this reason small particles in colloids and fine suspensions are set into Brownian movement visible under the microscope or the ultramicroscope, owing to the fluctuations of the surrounding molecules, which are in irregular thermal motion. They are struck irregularly by the molecules bombarding them, and hence display a ceaseless zigzag motion that is, so to speak, a magnified image of molecular movement. As was stated first by Nernst, F. Exner, and others, the second principle is not an exceptional case, but all laws of physics are of a statistical nature.

The fundamental limit of physical determinism was reached in quantum theory. If we come to the elementary physical events, we are faced with two basic and inter-related facts. First, while macrophysical processes appear to be continuous, that is, mechanical energy, light, electricity, etc., can be transferred in any arbitrarily chosen amount, a limit is reached in elementary physical events. If, for example, an atom emits or absorbs light, this does not happen in any small amount, but it happens in elementary units; either the full amount of energy of a light quantum is emitted or absorbed, or nothing at all. Secondly, it is impossible, in principle, to denote elementary physical events in a deterministic way. A simple example is radioactive decay. In this the nucleus of radium emits an α-particle by a sort of explosive process and thus changes into radon. If we have, say, a milligram of radium, we can state for certain that in about 1590 years this aggregate of an enormous number of atoms will have decayed to a half. We cannot say, however, whether an individual atom will decay in the next minute or perhaps only after thousands of years. Even if we could determine the state of the nucleus at any given moment, it would be fundamentally impossible to foretell when its decay will occur. For if there would be a further causal determination, the decay should depend on age, on external factors such as temperature, or the like. This is not the case. The only statement we can make is that, given a number of atoms, the same percentage will decay per unit time.

According to the testimony of modern physics, we can, therefore, make univocal forecasts with respect to macro-physical events, that is, events which concern a virtually infinite number of elementary physical units, for in such events the statistical fluctuations are evened out. For this reason the laws of classical physics, those of mechanics, for example, have apparently a strictly causal or deterministic character. With respect to microphysical events, however, which concern individual elementary physical units, no univocal forecast can be given, but only a statistical probability. The laws valid here determine only the average behaviour of numerous elementary particles; the behaviour of an individual unit can be prophesied only by indicating a certain probability.

This is, in a bare outline, the contrast between determinism, that is, the strictly causal laws of classical physics, and indeterminism, the statistical laws of modern physics. What is the meaning of this fundamental change for biology?

A living organism consists of an inconceivably large

number of molecules and atoms, the order of which is about a quadrillion. Thus, it is obvious that for the majority of biological phenomena, such as metabolism, growth, morphogenesis, most phenomena of irritability, etc., deterministic laws in the fashion of the laws of classical physics must apply.

However, there are certain biological phenomena which may be exceptions. The author, back in 1927, and even before the formulation of the Heisenberg relation, which forms the basis of modern indeterminism in physics, had raised the question of the " bearing on biology of the revolutions in physics." In 1932 he argued that the possibility should be considered " that microphysical events in the organism may be transmitted to wider regions of the system and thus lead to the circumvention of physico-statistical probability." The same idea was developed by the physicist Pascual Jordan to an " amplifier theory of the organism," according to which microphysical events in controlling centres, for instance, genes, are magnified in organic systems to macro-effects. Similar ideas were advanced by physicists such as Bohr, Schrödinger, and others. Both kinds of physical indeterminacy can be taken into consideration here, the " classical " fluctuations according to kinetic theory as well as quantum-physical indeterminacy. Research shows that in certain biological processes microphysical events are actually decisive.[1]

The first and most important field is radiation genetics, elaborated by Timoféeff-Ressovsky and his co-workers, that is, the induction of mutations by rays of short wave length, like X-rays, radium, or neutron rays. These investigations led to the " hit theory " of mutation. The action of radiation on biological objects may be compared

[1] For " quantum biology " compare for example P. Jordan, *Die Physik und das Geheimnis des organischen Lebens.* 2nd edition, Braunschweig, 1947; E. Schrödinger, *What is Life?* Cambridge, 1945; N. W. Timoféeff-Ressovsky, *Experimentelle Mutationsforschung in der Verbungslehre.* Dresden and Leipzig, 1937; N. W. Timoféeff and K. G. Zimmer, *Biophysik,* Vol. 1: *Das Trefferprinzip in der Biologie.* Leipzig, 1947.

M

to a bombardment of the sensitive material. Now radiation consists of quanta as discontinuous units of energy. As can be shown by mathematical analysis of the experiments, one single quantum hit into the sensitive zone of a gene suffices to cause a mutation. Therefore, the induction of mutations is subject to the statistical laws of microphysics; these microphysical events, however, are amplified by the organization of the living system to a macro-effect, and thus a mutation induced by radiation will become manifest at the macrophysical level, say, in a change of the shape of wings or the colour of eyes in a descendant of the fruit fly submitted to the treatment.

A second field of biology in which control by microphysical events seems to be proved is that of the destruction of micro-organisms. For example, if a culture of bacteria is subjected to radiation or to a disinfectant, the cells are killed at different times. The simplest explanation would be that the individual cells have a different susceptibility to the noxious agent. In this case the susceptibility, and hence the times of death of the cells, should follow a variation curve, the greatest number of individuals showing medium susceptibility and time of survival, and a smaller number exhibiting very high or low susceptibility and correspondingly short or long survival times. Actually, however, the death curve of the bacteria is an exponential, comparable to that of the decay of radium atoms (p. 164). That is, the number of cells killed per unit time is simply proportional to the number of cells present. This shows that destruction occurs as a matter of chance, due to a " hit " into a sensitive centre.

A third field in which, perhaps, microphysical phenomena should be taken into account was suggested by von Bertalanffy (1937). If animals are brought under the influence of a directed stimulus, a light-source, for example, orientation movements result in which the animal, according to whether it reacts to the stimulus positively or negatively, moves towards the light-source or away from it. Animals in a homogeneous environment, however, in darkness or diffuse illumination, for example, usually show "spontaneous" movements, in which both direction and speed change irregularly without recognizable external stimulus. It seems alluring to explain this behaviour by assuming that in the absence of optical marks for orientation, the animal is not able to transmit impulses of equal intensity to the locomotor apparatus on both sides. The greater the difference between the impulses that go to the right and left halves of the body, the more will the path bend towards the side which is less active at a given moment. The irregularity of the changes of direction in the run of the same animal shows that the deviations from the straight line cannot be due to persisting morphological conditions such as lateral asymmetries; they must depend on changing physiological conditions in the nervous system. It is known, moreover, that even in unstimulated nerve centres irregular volleys of action currents occur which indicate spontaneous excitations. It can therefore be imagined that spontaneous discharges occur in the central nervous system owing to the incessant processes of metabolism. Since the number of these discharges is small, they will be unequally distributed in both lateral halves, and hence cause irregular changes in the run. If, on the other hand, an external stimulus, a source of light for example, is applied, it establishes a definitely stronger excitation in one half, and leads thereby to movement in a straight course. But even in this case, changes in the direction and speed of movement can be observed that are not accountable to external stimuli, and are perhaps caused by spontaneous fluctuations in the excitation of the nerve centres.

Thus it is very likely that microphysical events must be taken into account in certain biological fields. We do not believe, however, that the solution of the "problem of life" can be found in this way or in any other single feature, for that matter. Above all, we must beware of the frequently advocated analogy between physical indeterminacy and free will, problems which lie at absolutely different levels. Physical indeterminism states that it is only the statistical behaviour of collectives, not individual events, which can be stated in physical laws. On the other hand, the conception of free will in ethics does not mean that events are statistically at random, but rather that they are subject to a norm; its very meaning is that action in a given situation is not fortuituous but determined by a moral principle. The supposition that free will interferes in the gaps left by physical causality is in the same line as the vitalistic assumption of a control of material events by entelechy. We cannot refute the latter assumption by strictly demonstrating that there is no entelechy active in the organic world. For we cannot make Laplacean forecasts for the organism, we do not completely survey its physical configuration, and it is therefore always possible—even if classical determinism is upheld—to fill the gaps in our knowledge by assuming the " intervention " of a vitalistic factor. Similarly, on the basis of scientific data the conception cannot be refuted that a free will determines a microphysical event that is not determined by the statistical laws of physics. Both assumptions, however, are a *metabasis eis allo genos*, since physical events and mental experience lie on two different levels of reality. Physics is only concerned with objective phenomena and their laws; the interference of mental factors into physical events—no matter whether it means direction of material atoms in the classical or intervention in micro-events in the modern interpretation—is outside the limits of physical theory. A presentation of the relation of physics and psyche, nature and mind, from the organismic standpoint will be tried in a later context (following volume).

4. *The Methodological and the Metaphysical Problem*

The antithesis between biological mechanism and vitalism is double-rooted. It is both a methodological and a metaphysical problem.

The methodological problem concerns the question of

what principles and laws are to be applied in the explanation of biological phenomena. This has been discussed in detail in the last sections. This discussion is not unnecessary, for the relation between physics and biology, inanimate and animate nature, belongs to the fundamental problems of scientific thought, which every era has to answer in its way. However, it seems advisable to repeat a warning given elsewhere (von Bertalanffy, 1932) :

" The dispute about whether biological laws will in the last analysis be physical or not—a dispute which formed a main part of theoretical biology— seems a rather fruitless business ; for, as the saying goes, you cannot hang anybody before you have caught him. What the organismic conception actually strives to is much more essential than a dubious negative prophecy for the future, namely, a positive research programme for the present. It points to the fact that the physico-chemical explanation of isolated processes in the organism, as it was almost solely used hitherto, contributes nothing to the insight into the law of order that turns these processes into biological phenomena, and that it is the essential task of biology to find the system-laws of the organism, a task hitherto hardly contemplated by ' mechanistic ' biology."

On the other side is the metaphysical question, namely, whether every event in the world, including biological events, is determined univocally by the ultimate physical units and the forces operating between them according to the laws of nature, or whether in the realm of the living other elements of reality, ultimately of a psychical nature, operate to direct the play of those particles. This question is meaningless. For both conceptions are based upon the mechanistic conception of classical physics, and in the light of modern physics and epistemology none of the notions used is consistent. The question whether the world process is determined " univocally " by the ultimate physical units becomes void, that is a statement that can neither be proved nor disproved, if the ultimate physical events are not accessible, for reasons of principle, to complete determination. The " laws of nature " do not represent any more the manifestation of forces that, no matter whether thought to be causal or final agents, are an anthropomorphism : causal forces being modelled after the image of the push I give a thing, final forces after our own purposeful action. In modern physics, the laws of nature appear as symbolic representations of formal relations between phenomena. In the last resort, they are statistical statements about certain collectives, not factors causing the course of events. Finally, ultimate physical units are not " material atoms " as a metaphysical reality, but they can be described only formally by mathematical expressions, and physics tells nothing about their " inner nature." Thus the antithesis between metaphysical mechanism and vitalism becomes a pseudoproblem because its premise, namely, the metaphysical dualism between inert matter as a metaphysical reality and a directing psychoid agent, is based upon a world-view of physics that no longer exists.

It has sometimes been said that the organismic conception is not a true solution of the mechanism-vitalism controversy. In fact, it does not fit into the usual alternative. The " mechanist," who would like to resolve the phenomena of life into physics and chemistry, finds a disturbing reference to laws and patterns that transcend physics and chemistry and so seem vitalistic to him. The " vitalist," on the other hand, regards those specific biological laws as mechanistic because they emerge from physico-chemical laws and are, in their logical structure, not different from the latter. Actually, however, it is the kernel of the organismic conception that it overcomes the mechanism-vitalism alternative at a higher level. The specific laws of the living state, which are denied by the mechanist and are considered by the vitalist as lying outside science, here become a problem accessible to scientific research.

Therefore a new methodical attitude is indicated. The organismic method is to find exactly formulated laws for organic systems as a whole. The word "exact" is taken seriously and meant in the sense that it is used in physics. As compared with the investigation of isolated phenomena, which, however, will always be necessary and must be propounded as far as possible, this is a new maxim of research which has already warranted its efficacy in many fields.

As far as the philosophical problem is concerned, the organismic conception says everything the scientist is entitled to enounce. He makes no statements about the "nature" of things and hence also about the question of an "essential" difference between the living and the non-living. Actually, the mechanism-vitalism alternative is not a controversy between two scientific explanations, one of which tries to explain the phenomena of life by physico-chemical, and the other by specific laws of some other type. The real difference is one between scientific explanation and anthropomorphic "understanding." Science is limited to the description and explanation of objective phenomena, "explanation" meaning fitting of these phenomena into a theoretical system (p. 161). The task of the vitalist is different : what he tries to do is to understand the "inner nature" of things, interpreting it after the image of our own inner experience. In the metaphysical sphere the psychic interpretation of reality may find the place that science cannot grant it. Then it is no more a scientific explanation but rather a spirited expression of mythical sentiment, a metaphor and allegory of the Unspeakable. It is the peculiar hybrid position between science and poetry by which vitalism is sapped. Seeking organic wholeness not in objective nature, but in a transcendent principle of life, it yields no basis for biological theory. On the other hand, it makes shallow the metaphysical intuition by rationalizing it and trying to introduce it into science as a causal agent. That mythical and metaphysical view of reality may be true or illusive—it is not a matter of science.

5. *Science—a Hierarchy of Statistics*

All laws of nature are of a statistical nature. They are statements about the average behaviour of collectives. Science as a whole appears as a hierarchy of statistics.

At the first level is the statistics of microphysics. In the realm of elementary physical events, a deterministic treatment is impossible in principle. As was said above, microphysical events also intervene in certain biological phenomena.

A second level is constituted by the laws of macrophysics, that is, of those phenomena where a great number of elementary physical units is concerned. These laws are also essentially statistical. Since, however, the statistical fluctuations are levelled out by the law of great numbers, macrophysical laws have an apparently deterministic character. As compared with the statistics of elementary physical events, the laws of macrophysics are at a higher level. For example, the laws of macromechanics or of hydrodynamics no longer take account of elementary physical events, for the simple reason that we cannot, and need not, run after every molecule but are satisfied with a lump treatment of the system.

A still higher level is represented by the biological realm. As explained above (p. 155), we can, on the one hand, isolate single processes and define them in terms of physics and chemistry. On the other hand, we can state overall laws for the biological system as a whole, foregoing determination of the individual physico-chemical processes comprised.

Finally, there are the laws that apply to supra-individual units of life. We can, for example, state laws for the growth of populations within a biocœnosis (p. 52) or for the incidence of death within a human population. Laws of this kind are the basis for insurance statistics, and hence are of great practical and commercial importance. The units taken into account here are individual organisms, and it is impossible, as well as unnecessary, to take into account in these laws the physiological or physico-chemical processes concerned.

Thus on the different biological levels laws can be established that are exact and quantitative and constitute a hypothetico-deductive system. In these respects they are comparable to the laws of physics, but, in contrast to the latter, they reckon with units of a higher level.

In this hierarchy of statistics we find a remarkable phenomenon which we can describe as an increase in degree of freedom.

For example, ordinary chemical compounds are defined by structural formulæ that determine univocally the number of atoms or radicals combined. This is true even for complex organic molecules. Coming to macromolecular compounds, however (p. 26), statistical values replace the rigid formulæ. One can only say that, for example, on the average three hundred sugar residues are combined in one main-valency chain, and on the average some sixty main-valency chains in a micella of cellulose.

The same is true for spatial arrangement. The crystals of minerals are three-dimensional lattices. In the organic realm, on the contrary, "mesoforms," that is, molecular arrangements in two or one dimension only, play a decisive part. They form, for example, the numerous fibrillar structures which are of paramount importance in the construction of the cell and the organism, the supporting tissues, muscles, nerves, etc.; in these fibrillæ, thread-like molecules are ordered parallel to the axis but unordered in the other directions.

Corresponding considerations apply to chemical processes. Here too, with increasing complexity we find an increase in the degree of freedom. Chemical processes in the organism are carried through by catalysis, acceleration of reactions that would otherwise be slow or would not take place. Simple catalyses, such as, for example, the combination of a mixture of hydrogen and oxygen to form water brought about by means of spongy platinum, can take place in one way only. However, in the more complicated catalytic actions in use in the chemical industry and, especially, in those occurring in living organisms, there are several possible ways of reaction. For example, with appropriate temperatures and pressures, carbon monoxide and hydrogen may combine to give either methane or methyl alcohol, or higher alcohols, or liquid hydrocarbons. The use of a nickel catalyst produces only methane, a zinc oxide-chromiumoxide catalyst practically pure methyl alcohol, and so on (Mittasch). In systems of this kind there are several ways of reaction that are thermodynamically possible, and which one will take place depends on the catalyst used. The skill of the industrial chemist consists of selecting catalyser systems suitable for definite purposes. In a similar way the variety of possible ways in which chemical reactions can take place in the organism is an essential basis for physiological processes.

In a crystal the external form is the expression of a molecular arrangement determined by the crystal lattice. Since, for example, the crystal lattice of sodium chloride represents a miniature cube, the macroscopic crystal also has a cubic shape. With organic forms it is very different. They are determined as a whole, with considerable variation in the arrangement of their elements. We may recall, for example, the shape of a pileus, which is predestined and characteristic for the species, and is built up from hyphæ that grow in all directions and show no law in their arrangement. We can express the distinction between inorganic and organic shapes in the following way. In the former the structure, that is, the law of internal arrangement, is constant, and the form, or outer shape, is the expression of the structure and can vary. In this sense the structure of a crystal, for example, is determined by the lattice, the distortions found in most crystals being non-essential. In contrast, in living systems the structure is variable, whereas the shape is determined. The latter appears as a mould, so to speak, that is filled by cells varying to a wide extent in number and arrangement. It is interesting to note that the high-molecular compounds are intermediates also in this respect. In proteins a common " mould " can be filled by different amino acids. For example, the keratin of hair, the contractile myosin of muscle fibres, and the coagulating fibrinogen of the blood, though chemically and physically different, have the same molecular pattern (Astbury). The chemotherapeutic action of sulphonamides probably depends on the structural similarity between the molecules of the sulphonamides and those of bacterial growth-substances; therefore the former can displace the latter, and so inhibit growth and reproduction of the bacteria.

Again, in a different way the increase of the degrees of freedom is shown in equifinality. Whereas the development of closed systems to the final state is determined by the initial conditions, the same end state can be reached in open systems in different ways.

Finally, we find a comparable phenomenon in phylogenetic and historical evolution. Certain overall laws seem to be fixed; their realization in the special case, however, depends on chance: on the one hand, on the appearance of suitable mutations, and on the other hand, of dominating personalities.

Thus, in the hierarchy of statistics the degree of freedom seems to increase as we move to higher levels; not in the sense of the indeterminacy of elementary physical events but rather in the sense that the process as a whole follows definite laws, but different possibilities are left open to the individual events.

CHAPTER SIX

THE UNITY OF SCIENCE

Aristippus philosophus Socraticus, naufragio cum eiectus ad Rhodiensium litus animadvertisset goemetrica schemata descripta, exclamavisse ad comites ita dicitur : Bene speremus, hominum enim vestigia video.—VITRUVIUS, De architectura.

> *In Xanadu did Kubla Khan*
> *A stately pleasure-dome decree,*
> *Where Alph, the sacred river, ran*
> *Through caverns, measureless to man*
> *Down to a sunless sea.*—COLERIDGE, Kubla Khan.

1. *Introduction*

IF we survey the various fields of modern science, we notice a dramatic and amazing evolution. Similar conceptions and principles have arisen in quite different realms, although this parallelism of ideas is the result of independent developments, and the workers in the individual fields are hardly aware of the common trend. Thus, the principles of wholeness, of organization, and of the dynamic conception of reality became apparent in all fields of science. We could enumerate still further common traits, such as the realization of the basically statistical character of the laws of nature and the intrinsic contradictoriness of reality. It appears that the description of reality in conceptual constructs cannot be achieved by a single scheme but rather by pairs of opposed and complementary conceptions. This is expressed in the complementarity principle of quantum theory (p. 180); in a different form complementarity probably also applies to the description of biological phenomena (p. 155). Another basic insight is that of the discontinuous nature of primary events, as opposed to the concept of continuity in classical physics. According to quantum theory, the ultimate units of reality are discontinuous and are not further divisible. Its counterpart in biology is the mutation theory, according to which evolution proceeds, not in continuous transitions but in discontinuous jumps. It is more than an historical accident that the quantum theory and the mutation theory, of which the latter stands in close relationship to the former (pp. 95, 165 f.), were founded in exactly the same year, 1900. Perhaps we are allowed also to add the all-or-nothing law in physiology, propounded at about the same time. According to this principle, the action, say, of a muscle or a sense organ does not increase continuously, but jump-fashion, because, with increasing intensity of the stimulus, new elements, each of which is functioning at its maximum, are brought into action.

2. *Physics*

Classical physics sought to resolve all processes in nature into a play of atoms, tiny particles moving in space according to the laws of mechanics, of push and pull. Modern physics has not only demonstrated directly the existence of the atoms; it has revealed their structure and conquered completely new fields in radioactivity, transmutation of the elements, and the release of atomic energy. These very developments, however, overthrew the mechanistic conception.

It was a first maxim of mechanistic physics that it is possible to resolve physical processes into separable local events. In opposition to this, the notion of wholeness appears necessary in modern physics. According to Heisenberg's uncertainty principle, it is not possible to determine simultaneously the position and momentum of an electron. In order to determine its position, the electron must be illuminated; but this means that a light quantum hits the electron, and so its momentum is changed. Therefore, the more exactly the position is determined, the less exactly is the momentum, and *vice versa*. From this there follows, first, the impossibility of a strict determinism in the microphysical realm (pp. 163 f.), since the uncertainty relation sets an insurmountable limit to a simultaneous determination of all measures necessary. Secondly, according to the Heisenberg relation, in physical micro-events the measuring instrument cannot in principle be separated from the entity measured. Thus a principle of wholeness appears in microphysics, in fact, in a still more radical sense than at the macrophysical level (p. 192). For in microphysics it is not simply a question that for understanding the whole the elements as well as the relations between them must be known; rather a further resolution becomes impossible, in principle, at the level of the elementary events, and they can be treated only as a whole.

Secondly, principles of organization appear most significantly in modern physics.[1] While the classical laws are ultimately laws of disorder, the central problems of modern physics and chemistry are problems of organization. As demonstrated by Boltzmann, causality works towards the destruction of order, since in the course of time thermal motion increasingly and irrevocably destroys all order that existed at the beginning. An atom, however, say a mercury atom, which consists of a nucleus and eighty planetary electrons, preserves its organization, upon which the system of spectral lines emitted, its chemical properties, etc., depend; and it does so in spite of the incessant bombardment to which it is subjected by the thermal agitation of the surrounding particles. The stability of the atom and the preservation of its organization in spite of the disturbances due to thermal motion are based upon the discontinuous nature of elementary physical events, as stated by quantum theory. The atom cannot stay in any state whatsoever, but can assume only discrete states with different quantum contents. If these states are denoted by the numbers 1, 2, 3, etc., then state 1 is the ground state with the least energy content, the state in which the atom normally exists; 2, 3, etc., are excited states that are reached in jumps if the necessary energy is provided. For this reason, disturbances that are too weak are ineffective, and thus the atom can remain stable for an unlimited time in spite of thermal motion. Only if temperature is increased, does it pass, by means

[1] A. March, *Natur und Erkenntnis*. Vienna, 1948.

of a quantum jump, into an excited state. Corresponding considerations apply to the molecule, the crystal, the solid state, and even the gene. The latter is a macromolecule of specific organization and high stability. Therefore, only in relatively rare cases, be it through a quantum hit in the case of induced mutations or through thermal fluctuations in the case of spontaneous mutations, will a transition into a new stable state and hence an hereditary variation take place. Herein lies the connection between quantum theory in physics and mutation theory in biology. The transition of a gene molecule into a new stable state occurs only by way of a jump, because the transfer of energy does not take place in any small quantities but is quantized. Biologically, this gives the interpretation that the transition from one race to another is not continuous, but also takes place in jumps (cf. p. 95).

A third basic change in modern physics is in the resolution of rigid structures into dynamics. Classical physics considered the atoms as solid bodies like tiny billiard balls. According to modern physics, they are minute planetary systems, with a nucleus as the central sun, consisting of positively charged and uncharged particles (protons and neutrons), encircled by negative electrons. At the same time matter appears as a process, as dynamics. The opposition of mass and force, matter and energy, plain in daily life and in classical physics, disappears on the microphysical level. An electron is not a tiny rigid body; it is a concentration of energy, a matter-wave or a wave-packet. For this reason, transformation of matter into energy and energy into matter is possible. A quantum of gamma radiation, that is, X-rays of high frequency, can convert into a twin pair of electrical particles, negative and positive, of an electron and a positron. Conversely, matter can be converted into radiation. The classical principles of the conservation of mass and of energy are united into the comprehensive conservation law of Einstein. Moreover, under certain conditions elementary physical units behave as corpuscles, under others, as undulations or waves. According to

Bohr's complementarity principle, corpuscle and wave are antithetic but equally necessary and supplementary conceptions of the same physical reality.

Wholeness, organization, dynamics—these general conceptions may be stated as characteristic of the modern, as opposed to the mechanistic, world-view of physics.

3. *Biology*

The movement of biological thought in recent decades towards " organismic conceptions," the significance of which is even more evident because it became to a large extent subconscious and anonymous, is no isolated phenomenon. Rather it is part of a general metamorphosis in our scientific conceptions.

We have considered the influence in biology of the mechanistic view in physics. Following the latter, the goal of biology was seen in the resolution of the phenomena of life into isolable parts and processes (pp. 10 f.). Thus the organism was conceived as a sum of cells and its function as a sum of cell actions. In the same way as physical events seemed to be ruled by the laws of chance, so the organization and function of the organism were regarded as products of random mutation and selection. This view corresponds, on the other hand, to economic trends and theories. In fact, Darwin generalized the Malthusian theory that the increase of human populations outruns the increase in their resources, and applied it to the entire living nature. The struggle for existence in the organic world is nothing else than free competition, advocated by the Manchester school at the beginning of the industrial era, and applied to biology. The utilitarian conception in biology was in line with the general ideology. The machine theory of life is the perfect expression of the *zeitgeist* of an epoch which, proud of its technological mastery of inanimate nature, also regarded living beings as machines.

Realization of the limits of the mechanistic conception led first to vitalism, which assumed the aggregate of parts and machine-structures to be controlled by purposive agents. In turn, the recognition of the inadequacy of both views led to organismic conceptions, which attempt to give a scientific meaning to the concept of wholeness. This is the common trend which we can find equally in biology, medicine, and psychology.

We have dealt in detail with the conceptions underlying modern biological thought and their effects in the different fields. There is, first, the concept of wholeness. Not only must the parts of the organism and the individual processes be considered, but also their mutual interactions and the laws governing the latter. These are manifest everywhere in the phenomena of regulation following disturbances, as well as in the normal functioning of the organism. Further, we have the concept of organization. The basic feature of the organic world is the tremendous hierarchy which extends from the molecules of organic compounds through self-multiplying biological units to cells and multicellular organisms, and finally to biological communities. New laws appear at each level of organization, and it is the task of biological research progressively to unveil them. Finally, there is the concept of dynamics. Living structures are not in being, but in becoming. They are the expression of a ceaseless stream of matter and energy, passing through the organism and forming it at the same time. The dynamic conception forms a basis for exact biological laws in many fields, and equally for understanding such phenomena which, like equifinality, have hitherto been regarded as mysteries that could not be explained in scientific terms.

Although similar views have been advanced by many writers in the last decades, the present author believes that he may claim, in his organismic conception developed since 1926, to have first consistently formulated the new attitude as a working hypothesis for biology. The fecundity of this conception can be seen in many conclusions then drawn, and verified and elaborated in subsequent research. It may be profitable, then, to summarize once again these developments.

N

Many scientists have accepted the organismic view, and it is interesting to see that some of them came from opposing camps. The biochemist Needham, for example, severely criticized the earlier holistic conceptions in biology, later to adopt the organismic conception. Biological theory is centred, as Needham (1932) says, on the problem of organization. While J. S. Haldane considered it a sufficient explanation of biological problems to refer to the organization of the systems in question, the organismic conception of von Bertalanffy and Woodger has shown the necessity of investigating what organization really is. Thus, organization is not an explanation, but is the most fascinating and difficult problem in biology. Recognition of this fact has nothing to do with vitalism. On the other hand, Alverdes (1933), working in the field of animal behaviour, first contended a vitalistic view and later accepted the organismic conception. The works of Alverdes (1936), Bavink (1929), Canella (1939), Gessner (1932, 1934), Tribiño (1946), and Ungerer (1941) may be mentioned as further presentations of the organismic conception. Critical discussions of the latter were given, from the mechanistic side, by Bünning (1932), Gross (1930), and M. Hartmann (1937); from the vitalistic, by Wenzl (1938); and from intermediate standpoints by Bleuler (1931), Burkamp (1930, 1936, 1938), and Linsbauer (1934). Similar views, some advanced independently, others in mutual interaction with our work, are found in the writings of Bavink (1944), Bizzari (1936), Brohmer (1935), Dürken (1937), von Frankenberg (1933), H. Jordan (1932), O. Köhler (1930), Needham (1936, 1937), von Neergard (1943), Oldekop (1930), Ritter and Bailey (1928), E. S. Russell (1931), Sapper (1930), Ungerer (1941), Wheeler (1929), Woodger (1929), Woltereck (1940), and others. The physicist Schrödinger (1946) also arrived independently at a conception similar to the organismic, "that the living matter—while it does not escape the physical laws as stated up to now—probably contains different physical laws unknown hitherto, which nevertheless, once they are known, will form just as integral a part of this science as the former." The work of Mittasch (1935, 1936, 1938) on biocatalysis and the hierarchy of causality in nature is also closely related to the organismic conception; so is that of the Marburg school of Alverdes (1937) on animal behaviour, H. Jordan's discussion (1941) of the basic principles of physiology, and Hirsch's views (1944) on dynamic histology. In the field of the physiology of developments, Dalcq (1941) denotes the organismic conception as that standing closest to his own. Without entering into a detailed discussion, it can be noted that the general trend of modern biology is in line with the organismic conception, and that this working hypothesis finds application in all fields of biology. Here only a survey of applications made by the author and his co-workers, and of closely related developments, will be given.

With regard to the problem of *vital organization*, the author has stated, in 1932, as a programme for future research :

" It seems a thoroughly arbitrary assumption, leaving the actual problem—the orderliness of the living processes in the organism—completely in the dark, that the hierarchy of physical structures should end with the micellæ of proteins, and that beyond this limit only the laws of disorder (i.e., the laws of probable distributions in solutions, as emanating from the second principle of thermodynamics) or the molar laws should apply; so that organization would be either a mere ' mixture ' or a rigid ' machine.' Rather it is probable that there is a continuous transition from micellar arrangements, the laws of which are partially known, to that even more non-rigid and dynamical state of order the laws of which are yet unknown, and which is called ' vital organization ' of the protoplasm and the cell. Certainly the living organization is not only ' non-rigid ' but also ' dynamic.' Here the problem of ' organization ' joins up with that of the ' steady state.' "

This challenge was met, to a then unexpected extent, by submicroscopic morphology of the protoplasm, as initiated by Frey-Wyssling. Zeiger (1943) has confirmed that a " dynamic " conception of protoplasmic organization (p. 34) is necessary. The conceptions on *Cell Theory* and its limitations as developed in our earlier work (1932, cf. pp. 38 ff.) correspond with the interesting work of Huzella on " intercellular organization " (1941). On the higher levels of organization, the dynamic conception overcomes the apparent opposition between *structure* and *function*, considering the organism as a hierarchy of processes going on with different speed. This conception was stated by von Bertalanffy and Benninghoff (1935, 1936, 1938; cf. pp. 134 ff.). The concept of *homology* has been redefined from the dynamic and organismic viewpoint (von Bertalanffy, 1934; see the following volume). The considerations on *biological individuality*, as given by von Natzmer (1935), correspond almost literally with ours.

Lugmayr (1947) has discussed the latter from the stand-point of Thomistic philosophy.

The ideas of organismic biology have also been found useful in ecology. In *forestry* Lemmel (1939) derived the principle of *dauerwald* from the organismic conception of the forest as a biocœnosis that maintains itself in the change of the individuals. This is an interesting example showing that the organismic conception is not only of theoretical value, but can also be applied to important practical and economic issues. Vanselow (1943) also states that the modern ideas in forestry correspond with organismic biology. The notion of *umwelt* (environment) within the system of general biology was defined by H. Weber (1938, 1939) on an organismic basis. When introducing this term von Uexküll has emphasized only one side of the relations between organism and environ-ment, namely, the reactions to sense-stimuli. His con-cept of *umwelt* is therefore limited to sense physiology and is, in fact, pseudo-psychological. According to Weber, we must, however, define the concept of en-vironment more broadly. It denotes the total system of influences acting upon the organism, a system that depends on the specific organization of the organism, and, at the same time, makes possible its maintenance. There-fore, *umwelt* includes not only the things that can act as stimuli, but also the whole complex of conditions neces-sary to the maintenance of the organism. On the other hand, the concept of *umwelt* reaches its limit in the sphere of human activities. The *umwelt* of animals depends on their physical organization. In the evolution of science, however, a progressive elimination of anthropomorphic traits takes place, that is, those very qualities and categories that depend on the specific organization of the human apparatus of perception are eliminated (von Bertalanffy, 1937). This viewpoint is similar to Gehlen's[1] criticism of the notion of *umwelt* as advocated by von Uexküll; he too declares this concept inapplicable to cultural activities of mankind. The problem of the

[1] A. Gehlen, *Der Mensch*. Berlin, 1940.

uniqueness of man has also been discussed by the present author (1948; see the following volume).

The theory of *open systems* leads to new problems and insights in the fields of physics, physical chemistry, bio-energetics, and physiology (see pp. 125 ff., 131 ff., and the following volume). The work of Prigogine and Wiame (1946), Prigogine (1947), Reiner and Spiegelman (1945), Skrabal (1947), and others has already been mentioned. Dehlinger and Wertz (1942) have applied the theory of open systems to *elementary biological units* (viruses, genes), considered as uni-dimensional crystals in a steady state; a more detailed model conception was indicated by von Bertalanffy (1944; cf. p. 30).

Dotterweich (1940) has given a comprehensive study on "biological equilibria," although interpreting this concept very broadly and thus including phenomena of a diverse nature. His conception remains, therefore, largely formal. He distinguishes three applications so far made of the concept of "biological equilibrium": (1) the morphological "law of balance of the organs" (Geoffroy St. Hilaire, Goethe); (2) the biocœnotic equilibrium (Escherich, Friederichs, Woltereck, and others); (3) the physiological conception of the organism as a dynamic equilibrium or steady state (von Bertalanffy). Of these notions, the last seems the basic one. For the "balance of organs" can be conceived as a steady state attained by the organism in the course of its allometric growth (p. 139). Biocœnotic equilibria represent steady states, not of physico-chemical entities, however, but on the higher level of supra-individual units. A quantitative theory of competition, regulation, dominance, and determination in morphogenesis, based on a generalized kinetics of open systems (some-what in the line of our "System Theory") and on the gradient principle, was developed by Spiegelman (1945).

The conception of the organism as an open system leads to *dynamic morphology* (von Bertalanffy, 1941), that is, the interpretation of organic forms as the result of an ordered flow of processes. This necessitates an integration of morphological and physiological methods and points of view, and paves the way towards quantitative laws of *metabolism, growth,* and *morphogenesis.* The problems treated by the author and his groups in this field were enumerated earlier (pp. 136 ff.); a more detailed survey will be given in the following volume. Klatt (1949) has given an important discussion of dynamic morphology. Being one of the first (1921) to apply quantitative methods in morphology, and to introduce the law which is now known as allometry, he comments on the significance, bearing, and limitations of quantitative analysis of organic forms.

The agreement between recent experimental results and the conceptions on the *function of the nervous system* as derived from the organismic view has already been indicated (p. 121).

Medical science followed a development closely similar to that of modern biology. Virchow's Cellular Pathology aimed at resolving disease into disturbances of the cells. He rejected concepts such as that of constitution, which became so important again in modern medicine, precisely with the motivation that they are based upon the conception of the organism as a whole, which, in his opinion, is wrong. However, the evolution of modern medicine towards an organismic view is apparent; endocrinology or the theory of human constitutions are examples of organismic medicine.

As a matter of fact, the organismic conception was welcomed in medicine as a "liberating achievement." According to von Neergard (1943), H. Zimmermann (1932) was probably the first to realize the bearing of modern biological conceptions on medical practice. As he says, " because of the obvious correspondence between the evolution of leading ideas in medicine and those in theoretical biology, a point is reached that can be regarded as of historical significance." The organismic conception seems " to come closest to the leading ideas and necessary postulates of modern medical science." In a later paper

(1935), Zimmermann criticizes the so-called "biological medicine" from the basis of organismic biology. In a comparative survey of the theoretical trends in modern medicine, Rothschuh (1936) refutes the mechanistic, vitalistic, and psycho-vitalistic theories, and appreciates the organismic conception as a sound theoretical basis for modern medicine. Clara's presentation (1940) of the problem of wholeness in medicine closely follows the statements given by von Bertalanffy (1937). The gynæcologist Seitz (1939) is close to the organismic conception when he propounds a "holistic view of vital processes, normal and pathological," with respect to the biological, physiological, and medical problems of regulation in growth, sex, and reproduction. In general, the close correspondence of our biological conceptions and those of leading physicians like Aschoff, Bethe, Bier, Brugsch, and others may be emphasized. In physical therapy the work of von Neergard (1943) is closely connected with the organismic conception. There is also a striking correspondence between the conceptions of dynamic morphology and the work on types of human constitutions by Conrad (1941), of Kretschmer's Marburg school, though the two lines of thought developed quite independently. The influence exerted by the organismic conception upon medicine deserves special attention, for medicine is a fine test of biological theories, which are faced here with clinical practice.

The organismic conception has also found application in the field of *psychology*. Thumb (1944) has outlined its significance for psychology and appraised the bearing of the ideas of dynamic equilibrium and steady state as model conceptions in this field. Principles which are similar to those in the biological realm are found in the psychological. In particular, the controversy between the notion of environment in the biological sense (von Uexküll, Weber) and its inapplicability to man (Gehlen) disappears when, as in considering morphogenesis from the viewpoint of dynamic morphology, the building-up of human *umwelt* is contemplated from the viewpoint of developmental laws. Just as the conceptions of dynamics and wholeness in biology have their parallel in psychological *gestalt* theory, the hierarchy of biological organization has its counterpart in the strata of personality (Rothacker, 1947). The organismic conception has also been applied in the psychiatric and sociological fields (Burrow, 1937; Syz, 1936). Behaviour is conceived as a pattern of intra-organic tensions, and for psychotherapy it is claimed that the neurotic should be regarded not as an isolated individual but as one embedded in a social unit. At the same time, another side of the above-mentioned uniqueness of man (pp. 184 f.) becomes apparent. Antagonistic and co-operative tendencies are to be found in the animal kingdom, but only in man do we find hate, crime, and social anarchy. They appear to be connected with affective tendencies adhering just to those semantic attitudes—ideation and speech— which raise man above all other creatures.

There are many applications of the organismic conception in *philosophy*, of which the following are known to us as developments of our doctrine. Lassen, from the Cassirer School (1931), dealt with the problems of physical acausality and of teleology in connection with the organismic conception. Fries (1936) used this conception as a basis of an inductive metaphysics. Ballauff (1940, cf. also 1943) gives a synthesis of von Bertalanffy's organismic conception and N. Hartmann's doctrine of stratification (*schichtengesetze*) according to which reality is considered a succession of superimposed strata, each with its own laws. Ballauff adopts our definition of organic systems by means of the principles of hierarchical order and maintenance in a steady state, characterizes autonomy in the organismic way that what is permanent in the organism is solely its specific law of order, and exposes the philosophical consequences of this conception.

As a philosophically important deduction of our theory we have already mentioned the new conception of *equifinality*, which gives a physical foundation to the notion of directiveness hitherto considered as meta-physical and vitalistic.

The ultimate generalization of the organismic conception is the creation of a *General System Theory* (von Bertalanffy, first in 1945 ; see pp. 199 ff. and following volume) as the basis of an exact and mathematical ontology and of the logical homologies of the general concepts in different sciences.

Thus it might be said that the organismic conception has proved fruitful in many fields, from special problems of biology up to general questions of human knowledge. The most convincing evidence for this conception is that it has been applied in fields so utterly different as physics, physical chemistry, anatomy, embryology, physiology, forestry, medicine, psychology, and philosophy ; and has shed light on many problems in all these realms.

4. *Psychology*

The modern development of psychology is of special interest because the first scientific approach to the problem of wholeness was made in this field. Just as biology approached corporeal phenomena, classical psychology tried to resolve mental life into isolated events, psychical atoms as it were. A visual perception, for example, was considered to be a sum of elementary sensations corresponding to excitations of single cells in the retina and corresponding cells in the visual area of the cortex. The inadequacy of this conception soon became evident, and thus psychology introduced controlling factors like " apperception " according to Wundt—an interpretation comparable to the assumption of vitalistic agents in biological phenomena. *Gestalt* psychology was the attempt to overcome this dilemma. According to von Ehrenfels (1890), *gestalten* are defined as psychical states or events the characteristic properties of which cannot be obtained by adding up their components (first Ehrenfels criterion). Thus, for example, a perceived geometrical figure, a melody, or an intelligent sentence are, respectively, more than a sum of points of a certain colour, of sensations of single notes, and of meanings of the individual words. Furthermore, the same shape may be presented in other colours and on other parts of the field of vision, the same melody in other keys, the same sense in different words. Thus the *gestalt* remains the same when its parts are changed. *Gestalten* are transposable (second Ehrenfels criterion).

Now classical theory interpreted order in mental life by means of structural mechanisms. A sense organ, the retina for example, is subjected to numerous stimuli. From each point of the retina local excitation is led *via* a fixed nerve path to a corresponding end-point in the visual centre of the brain, so that there corresponds to the mosaic of retinal elements a similar mosaic of neurones in the cortex. In the same vein, recognition, association, conditioned reflexes, and so on were explained by the establishment of nerve paths between the centres concerned during the process of learning.

By contrast, *gestalt* theory demonstrated that it is not possible to resolve perception into a mere sum of elementary sensations and underlying excitations. For example, we recognize a figure as a triangle, even if it is presented in different sizes and in different parts of the field of vision. The points of the retina which are stimulated are different, and correspondingly the processes of excitation travel in different nerve fibres and to other nerve cells of the visual centre. Nevertheless, excitation of different retinal cells, nerve fibres, and cells of the visual centre produces the same impression, namely, " triangle." Conversely, excitation of the same cells may produce different impressions if, for example, those retinal cells on which a triangle first fell are stimulated later by the figure of a circle.

Gestalten are psychical wholes developing according to dynamical laws. The most important principle is that of pregnance, i.e., the tendency of *gestalten* to assume the simplest possible or most " significant " forms. If, for example, for a fraction of a second, nine points in a circular arrangement are presented to the eye, and a tenth some-

what outside the circle, the point outside seems to wander
into the circumference of the circle in order to complete
the most significant *gestalt* possible. Or if figures presented momentarily show little gaps, movements to close
the gaps are seen, the free ends of the figures flash
together. If a stick is so arranged that its projection on
the retina goes through the blind spot, it is seen without
any gap, whereas if a man is so placed that the projection
of his hand falls on the blind spot of the observer's fixed
eye, the head is not seen. The reason is that only
gestalten representing definite geometrical figures can be
completed.

Thus perception is not a sum of isolated and mutually
independent sensations, but sensations form configura-
tional systems which are governed by dynamic principles.

Probably the theory of memory must also be re-shaped
in a similar way. The classical view was summative and
mechanistic. It was assumed that traces or " engrams "
of earlier excitations remained in little groups of ganglion
cells, stored as it were in a myriad of depots intercon-
nected by myriads upon myriads of nerve paths—an idea
that is obviously impracticable (R. Wahle). If, however, *gestalt* perception is a system process that, dynamic-
ally organized, spreads over a larger cortical area, then the
after-effect of excitation will not consist of isolated traces
left in individual cells but rather in a certain alteration of
a larger brain-field. Actually, experimental and clinical
experience indicates that with respect to memory the
brain does not act as a sum of cells or sharply circumscribed centres. Localized injuries in the cerebrum do
not destroy one single function only, but others are
always affected, and the more strongly the higher their
requirements in terms of brain-function. Thus another
conception arises as opposed to path theory. It can be
assumed that the cerebral process during the learning
period, when two connected stimuli were active, represents an integrate whole. Correspondingly, it will leave
a unitary trace. After the period of learning, a new
partial stimulus will lead to the revival of the trace as a

whole and thus produce association, recollection, or a conditioned reflex (von Bertalanffy, 1937).[1]

If perception is not a mosaic of single sensations, but perceived *gestalten* organize themselves according to dynamic laws, we must conclude further that the physiological events corresponding to form-perception are not bundles or sums of single excitations but are also wholes or "*gestalten.*" Starting from this consideration W. Köhler (1924) raised the question whether *gestalten* are restricted to the psychical sphere. He emphasized that physical systems, in general, are not mere sums but satisfy the Ehrenfels criteria. Thus, for example, the distribution of electric charges on a conductor cannot be obtained by the summation of the charges on its individual parts but is dependent on the whole system. Also, it is re-established after part of the charge has been removed. In general, states (for example, the distribution of charges on a conductor) and processes (for example, the distribution of steady currents in a system of conductors) in physical systems depend upon the conditions in all parts of the system. Therefore they are characterized as *gestalten*. Finally, the same point of view was applied by Köhler (1925) to biological problems. The fact that the processes in an organism are regulated according to the needs of the whole is the most striking characteristic of the phenomena of life. Even a complete physico-chemical knowledge of all the individual reactions involved would not allow of its full understanding. Mechanists believe that this order is provided by machine-like structures; but this explanation fails in the face of the phenomena of regulation. Vitalists, on the other hand, invoke supernatural agents; but the dependence of parts on the whole, as in Driesch's sea-urchin experiment, is not a vitalistic feature, but a general characteristic of *gestalten*. Every system to which the second principle of thermodynamics applies finally reaches a state of equilibrium, which is characterized by the de-

---

[1] A substantially similar view was recently propounded by H. Rohracher (*Lehrbuch der Psychologie*. 2nd edition, Vienna, 1948).

pendence of the state in any part on the state in the whole system. So, in contrast to mechanism explaining the orderliness of processes in the organism by means of pre-established machinery, and vitalism, appealing to super-natural agents, there is yet a third possibility, namely, dynamic regulation within an integrate system. In this way, physics, biology, and psychology are equally concerned with systems in which the order of processes results from dynamics. The fundamental principle is that of equilibrium or of pregnance. In physics, it appears as the tendency towards minimum conditions that indicate states of equilibrium. In biology the orderliness of processes and regulation after disturbances as found in the organism can be equally considered as a consequence of the tendency to establish states of equilibrium. In psychology mental events appear to be *gestalten*; on the other hand, the demonstration of *gestalten* in the physical realm allows the interpretation of the underlying physiological events as *gestalt* processes. *Gestalten* as experienced appear as the correlates of equilibrium distributions of cerebral processes of excitation that tend towards the simplest possible configurations.

Köhler's conceptions mark the introduction of the modern system concept of the organism. The principal objections against *gestalt* theory are twofold. The first is that, lacking experimental possibilities, it did not advance far beyond the programmatic claim that experienced *gestalten* correspond to *gestalt*-like processes of excitation in the brain. Hardly any attempt was made to define more closely the physiological excitation-*gestalten* and, further, the system-processes underlying biological wholeness at large. General notions, however, such as "equilibrium," *gestalt*, and the like, are not an explanation, as was early emphasized by Driesch. What is necessary is an exact statement of the systems and processes in question and of the laws governing them. To what extent this seems at present possible in biology has been indicated earlier in this book. A second objection concerns the general type of distributions that, according to *gestalt* theory, are to be assumed in biological and psychophysical processes. Köhler tries to explain organic regulation by the establishment of states of equilibrium according to the second law of thermodynamics. But this conception is inapplicable in principle to living organisms because they are not systems in thermodynamic equilibrium but open systems maintained in a steady state which is remote from true equilibrium. Therefore the theory of organismic regulation necessitates new principles, which will have to be derived from the theory of open systems.

In any case, there is an astonishing correspondence between the development of modern psychology and biology. A modern text-book such as W. Metzger's *gestalt* psychology (1941)[1] could be translated, so to speak, theorem by theorem, into organismic language. We are inclined to think that a General System Theory (pp. 199 ff.) will be useful as a regulative device, to establish, on the one hand, those general principles which are common to the different fields, and, on the other, to guard against unwarranted analogies.

5. *Philosophy*

The future historian of our times will note as a remarkable phenomenon that, since the time of the First World War, similar conceptions about nature, mind, life, and society arose independently, not only in different sciences but also in different countries. Everywhere we find the same leading motifs : the concepts of organization showing new characteristics and laws at each level, those of the dynamic nature of, and the antitheses within reality.

The father of every philosophy of dynamics was Heraclitus; his " everything flows " and " unity of opposites " were the first, profound, and mystical expression of that world-view which, in latter days, we try to express in the sober language of the physical and biological sciences. From Heraclitus, the ideological trend leads to the enigmatic figure of the Cardinal Nicholas of Cusa in the Italo-German renaissance. Cusa is the last

[1] W. Metzger, *Psychologie*. Braunschweig, 1941.

in the line of great medieval mystics, and the precursor of modern science. He overthrew the geocentric system of Antiquity and the Middle Ages and taught the infinity of the universe. Thus he was a harbinger of modern astronomy as well as of the enthusiastic philosophy of Giordano Bruno. He meditated on infinity, and thus initiated that evolution which finally led to Leibniz's invention of Calculus. His physical, geographical, and medical observations mark the dawn of modern science and of the great intellectual movement that runs from Galileo to our time. In Cusa's doctrine of the coincidence of opposites, the ancient theme rose again to be passed on to modern times. The idea that reality—God in Cusa's formulation—can be spoken of only in antithetical statements is, interpreted in modern terms, also the profoundest criticism of the symbolism of language that eventually finds its most subtle expression in the complementary, and equally necessary, notions of modern physics. This intellectual heritage is perpetuated in the gloomy mysticism of Jacob Böhme, as well as in the lucid mathematics and natural philosophy of Leibniz, and in the poetic vision of Goethe and Hölderlin.

Goethe, not only a poet but also a great naturalist, was the founder of morphology, the science of organic forms. He conceives basic ground-plans, creative ideas, so to speak, of nature the artist, in the multiformity of animals and plants. Thus the protean diversity of plant forms is considered to be variations of an ideal original plant, the basic element of which—the leaf—is metamorphosed in different ways. It would be superficial, however, to see in Goethe's world-view only this element of " idealistic morphology," which is rooted in Plato's doctrine of ideas. Behind the ideal forms lies the Heraclitean dynamics, finding its expression in Goethe's *Stirb und Werde* (Die and Become), and *Dauer im Wechsel* (Duration in Change). Still, behind the beauty of forms, lies the contradictoriness of reality, and this makes our thoughts and actions merely symbolical. Hence, " the fiery flight of our mind is satisfied by images and pictures," and what we are doing is, after all, symbolical, so that, as Goethe said to Eckermann, it finally matters little " whether one makes pots or jars." Again, the Heraclitean coincidence of opposites is the core of the philosophy of the tragic visionary Hölderlin. He anticipates the inner contradictions in Hellenism, as exposed later by Nietzsche and Bachofen, mirrors them in his own soul, and is broken by them.

On such illustrious ancestors can the philosophy of nature of our times look back. These manifold independent sources flow into a common stream of thought.

Developments in philosophy preceded those in psychology and biology. Thus, Nicolai Hartmann emphasized in 1912 the necessity of the system conception. Considering causality as though single causal trains would run parallel is inadequate. What is essential is interaction. In a system the forces balance each other, and therefore their co-existence leads to a relatively stable configuration that resists destruction. At the same time, every finite system is a member of a higher one and encompasses smaller systems. This inclusion is not merely a passive encapsulation but mutual interdependence. Certain actions of the system of lower order are operative in the integration of the higher system. Conversely, certain actions of the higher system co-determine the lower ones. Living beings represent the most complicated configurations of systems of forces. Interaction is essential in them; it integrates all partial processes to the whole and governs their co-operation by system laws. In his later work Hartmann has developed the theory of stratification of reality. In its different realms—the inorganic, organic, and mental—it displays even higher and more complex categories.

We have described how the new conceptions of biology and psychology have evolved in the German-speaking countries. It is a most remarkable phenomenon in the modern history of ideas that a parallel and independent development took place in other countries as well. As Woodger has so nicely put it, a future history of biology will probably include a chapter entitled " The Struggle for the Concept of Organism in the Early Twentieth Century." It will describe how this idea was neglected under the influence of Cartesian philosophy ; how a mechanistic metaphysics did not allow biology even to dream of organisms as anything other than swarms of tiny hard corpuscles ; how the first appearance of the concept of organism at the turn of the century was frustrated by an inappropriate formulation, Driesch merely replacing the absurd notion of a machine without a mechanic by that of a metaphysical engineer ; finally, how the concept of organism was first taken seriously not by biologists but by some philosophers and mathematical physicists.

As Driesch had done in Germany, the English physiologist J. B. S. Haldane rejected the machine-theory of life. He saw in co-ordinated self-preservation the essential of life, and regarded this as not describable, in principle, in physico-chemical terms. Like the concept of *gestalt* in Germany, that of the organism was extended in England to include inanimate systems. According to Lloyd Morgan, the characteristic feature of an organism is that its parts owe their characteristic properties to the whole, losing them therefore after destruction of the latter. What Morgan has called " emergent " and " resultant " evolution, corresponds to the terms of *gestalt* and sum in German literature. Thus every level—electron, atom, molecule, colloidal unit, cell, tissue, organ, multicellular organism, and society of organisms—acquires, in emergent evolution, new characteristics that surpass those of the subordinate systems.

The " Organic Mechanism " of the mathematician Whitehead surpassed the conceptions of both the blind play of molecules and vitalism. All true entities are " organisms " in which the plan of the whole influences the characteristics of the subordinate systems. This principle is quite general, and is no prerogative of living bodies. In modern physics the atom has become an organism. Through the transformation of physical con-

o

cepts science touches on an aspect that is neither purely physical nor purely biological—it becomes the study of organisms. Biology is the study of larger organisms, physics that of smaller ones.

Haldane's doctrine was followed by the holism of Smuts and Meyer-Abich. According to this, biological laws are more general than physical ones. If, therefore, mathematical expressions could be given for biological phenomena, then, after elimination of the characteristically biological parameters, a simplified formula should be left which holds equally for both the living and the non-living, and which is identical with the physical law in question. However, there is, at present, no example where this procedure of a " simplifying deduction " of physical laws from biological ones (and biological from psychological) could actually be demonstrated. Thus, holism is a philosophical speculation, hardly supported by any facts in our present knowledge.

From the philosophy of Hegel, on the one hand, and the economic theories of Marx and Engels, on the other, originates Russian dialectical materialism. Its principles have been stated as follows : First, nature is not an aggregate of separate units but rather an organic whole, coherent and interacting. Secondly, nature is not a state of rest and immobility but one of incessant movement and evolution. Thirdly, in the process of evolution, jumps appear, governed by laws of nature, at the points of transition from one level of organization to a higher one, quantitative changes transmuting into qualitative differences. Fourthly, inner contradictions are dialectically immanent to natural phenomena, so that the process of evolution takes place in the form of a struggle of antithetical tendencies.

Of course, it would be nonsensical to blur the profound ideological differences and contrasts in all these movements and we shall refrain from any judgement on their merits. But these fundamental antagonisms make the " unity of opposites " all the more striking. The fact that from absolutely different and even diametrically opposed starting-points, from the most varied fields of scientific research, from idealistic and materialistic philosophies, in different countries and social environments, essentially similar conceptions have evolved, shows their intrinsic necessity. It can mean nothing else than that these common general concepts are essentially true and unavoidable.

6. *General System Theory*

From the statements we have made, a stupendous perspective emerges, a vista towards a hitherto unsuspected unity of the conception of the world. Similar general principles have evolved everywhere, whether we are dealing with inanimate things, organisms, mental or social processes. What is the origin of these correspondences?

We answer this question by the *claim for a new realm of science*, which we call General System Theory. It is a logico-mathematical field, the subject matter of which is the formulation and derivation of those principles which hold for systems in general. A " system " can be defined as a complex of elements standing in interaction. There are general principles holding for systems, irrespective of the nature of the component elements and of the relations or forces between them. From the fact that all the fields mentioned are sciences concerned with systems follows the structural conformity or " logical homology " of laws in different realms.

The principles that hold for systems in general can be defined in mathematical language. A more elaborate treatment will be given in the following volume. It will be seen then that notions such as wholeness and sum, progressive mechanization, centralization, leading parts, hierarchical order, individuality, finality, equifinality, etc., can be derived from a general definition of systems; notions that hitherto have often been conceived in a vague, anthropomorphic, or metaphysical way, but actually are consequences of formal characteristics of systems, or of certain system conditions.

O 2

The significance of a general system theory lies in various directions. At first, we may distinguish various levels in the description of phenomena. The first is represented by mere *analogies*, that is, superficial similarities in phenomena that correspond neither in the factors operating in nor in the laws applying to them. An example are the *simulacra vitæ*, much in vogue in biology at the beginning of the twentieth century, as, for instance, the comparison of osmotic " cells " with organisms they resemble. A second level is given in logical *homologies*. Here phenomena differ in the causal factors involved, but are governed by structurally identical laws. For example, the flowing of fluids and heat conduction are expressed mathematically by the same law, though the physicist knows, of course, that there is no " heat flow " but heat conduction is based upon the imparting of molecular movements. Finally, the third level is *explanation* in the proper sense, that is, the statement of the conditions and forces present in the individual case, and of the laws following therefrom. Analogies are scientifically worthless. Homologies, however, frequently provide very useful models and are widely used in this way in physics.

General system theory can serve therefore as a tool to distinguish analogies from homologies, to lead to legitimate conceptual models and transfer of laws from one realm to another, and, on the other hand, to prevent deceptive and inadmissible analogies and consecutive erroneous conclusions. In sciences that are not within the framework of physico-chemical laws, such as demography and sociology, as well as wide fields in biology, nevertheless, exact laws can be stated if suitable model-conceptions are chosen. Logical homologies result from general system characters, and this is the reason why structurally similar principles appear in different fields, and so give rise to a parallel evolution in different sciences.

General System Theory sets well-defined problems. Thus, for example, Volterra has shown that a demographical or population dynamics can be developed that is homologous with mechanical dynamics. A principle of least action appears in very different fields : in mechanics, in physical chemistry as Le Chatelier's principle (which, according to Prigogine, also holds in open systems), in electricity as Lenz's law, in the theory of populations according to Volterra, and so on. Again, relaxation oscillations (p. 141) appear in certain physical systems, and in many biological and demographical phenomena as well. A general theory of periodicity is a desideratum in various fields. Thus, the attempt will have to be made to extend principles like that of least action, of the conditions of stationary and periodic solutions (equilibria and rhythmic changes), for the existence of steady states, and so on, in a way which is generalized with respect to physics and so applicable to systems of any kind.

From the logico-mathematical standpoint, the position of General System Theory is similar to that of the theory of probability, which is purely formal in itself, but can be applied to very different fields, such as theory of heat, biology, practical statistics, and so on.

In philosophy general system theory may replace the doctrine known as " ontology " or " theory of categories " by an exact system of general principles. In fact, those very characteristics of knowledge and reality which have been stated by N. Hartmann under that title can be developed here in a mathematical form.

In this sense, general system theory may be considered as a step towards that *Mathesis universalis* which Leibniz dreamed of, a comprehensive semantic system including the various sciences. Perhaps it can be said that in the modern dynamic conception a theory of systems may play a role similar to that of Aristotelian logic in Antiquity. For the latter, classification was the basic attitude, and thus the doctrine of the relation of universals in their subordination and superordination appeared as the basic scientific organon. In modern science, dynamic interaction is the basic problem in all fields, and its general principles will have to be formulated in General System Theory.

7. *Finale*

The evolution of science is not a movement in an intellectual vacuum; rather it is both an expression and a driving force of the historical process. We have seen how the mechanistic view projected through all fields of cultural activity. Its basic conceptions of strict causality, of the summative and random character of natural events, of the aloofness of the ultimate elements of reality, governed not only physical theory but also the analytical, summative, and machine-theoretical viewpoints of biology, the atomism of classical psychology, and the sociological *bellum omnium contra omnes*. The acceptance of living beings as machines, the domination of the modern world by technology, and the mechanization of mankind are but the extension and practical application of the mechanistic conception of physics.

The recent evolution in science signifies a general change in the intellectual structure which may well be set beside the great revolutions in human thought. " The philosophical key-position of theoretical biology proves to be a second Copernican revolution in the history of our civilization, as von Bertalanffy once called it " (Thumb). In fact, the ideas to which modern scientific developments have led—wholeness, dynamics, organization into ever higher units—are most significantly shown in the living world. We may hope that these intellectual developments herald a new epoch which mankind is entering through the formidable crisis of our time—if it does not lead to general destruction. For spiritual revolutions always precede material developments. Thus, the theoretical conception of a mechanistic world, initiated by Descartes in the seventeenth century, was the precursor of the technicalization of life which has reached its zenith in our epoch. Similarly, we may perhaps consider the new scientific conceptions as harbingers of future developments. Not only for the poet but also for every creative work Hölderlin's splendid sentence is true:

" Er fliegt, der kühne Geist, wie Adler den Gewittern, weissagend kommenden Göttern voraus." [1]

There is still a final question that we have to answer. We have pinned down biology, in the organismic conception, at the level of pure science. We claimed that the phenomena of life are accessible to exact laws, though we may still be far from having reached this goal. We emphasized that any intervention of vitalistic agents in the observable, which forms the only subject-matter of science, must be rejected. The question then arises : Does this mean a bleak materialism, a soulless and godless nature ?

A glance at the most exact of the sciences answers this question. In a sweeping synthesis physics has come to a world view that comprehends reality from the incomprehensibly small units in the field of quanta up to the incomprehensibly large systems of galaxies. This control of nature, conceptual in physical theory and practical in technology, rests on the fact that phenomena are caught in a cobweb of logico-mathematical relations—which we call laws of nature. It is the triumph of modern physics that this fabric of laws of nature has achieved a universality and objectiveness never before reached. The trivial fact that the technological control of nature has been possible by means of those laws shows that they correspond in a wide degree to reality.

However, a certain resignation goes hand in hand with these achievements. In contrast to the self-assertion of former times, physics has realized that its task is the description of phenomena within a system of formal relations. It no longer expects to grasp the core of reality. Whereas earlier physics had thought that it has found the ultimate essence in tiny hard bodies, the statements of modern physics are different. Matter is resolved into some oscillatory process—but oscillation means only a periodic change of some magnitude, the ultimate nature of which remains undecided.

[1] " Like the eagle before the thunderstorm forboding flies the bold spirit before coming gods."

The physicist does not answer the question what an electron really " is." His most penetrating insight only states the laws that are characteristic of the entity called an " electron." Likewise, no answer may be expected from the biologist to the question of what life may be in its " intimate essence." Even with advancing knowledge, he too will only be better able to state what laws characterize, and apply to, the phenomenon facing us as the living organism.

Factors inaccessible to objective investigation must not intrude into the laws which can be stated for the observable. On an essentially different level lies a metaphysics trying to gain an intuitive knowledge of reality. We are not only scientific intellects, we are also human beings. To express in momentous symbols the core of reality, that is what myth, poetry, and philosophy are trying.

If, however, we aspire to grasp living nature in a brief sentence, it seems that this is given in a favourite expression of Goethe's. *Dauer im Wechsel* (Duration in Change) it is called in a profound poem. And the river that seemed the simile of life to Heraclitus, ever changing in its waves and yet persisting in its flow, also gives final knowledge to Goethe-Faust. Incapable of looking at the sun of reality, he and the scientific mind rest content with a great metaphor, holding, however, inexhaustible powers of life and thought :

> Behind me, therefore, let the sun be glowing!
> The cataract, between the crags deep-riven,
> I thus behold with rapture ever-glowing.
> Yet how superb, across the. tumult braided,
> The painted rainbow's changeful life is bending.
> Consider, and 'tis easy understanding,
> Life is not light, but the refracted colour.[1]

---

[1] Goethe, Faust II, B. Taylor's translation.
So bleibe denn die Sonne mir im Rücken ! Der Wassersturz, von Fels zu Felsen brausend, Ihn schau' ich an mit wachsendem Entzücken. Allein wie herrlich, diesem Strom entspriessend, Wölbt sich des bunten Bogens Wechseldauer. Ihm sinne nach, und du erkennst genauer : Am farbigen Abglanz haben wir das Leben.

REFERENCES

*1. Publications Pertinent to the Foundation of
the Organismic Conception (extract)*

BERTALANFFY, L. v.: *Kritische Theorie der Formbildung.* Berlin, 1928.
Nikolaus von Kues. München, 1928.
Lebenswissenschaft und Bildung. Erfurt, 1930.
Theoretische Biologie. I. Band: *Allgemeine Theorie, Physikochemie,
Aufbau und Entwicklung des Organismus.* Berlin, 1932.—II. Band:
Stoffwechsel, Wachstum. Berlin, 1942. 2nd ed., Bern, 1951.
Modern Theories of Development. Translated by J. H. WOODGER.
Oxford, 1933.
Teoria del Desarrollo Biologico (Spanish). 2 vol. La Plata, 1934.
Das Gefüge des Lebens. Leipzig, 1937.
Vom Molekül zur Organismenwelt. Grundfragen der modernen
Biologie. 2nd ed., Potsdam, 1948.
Biologie und Medizin. Wien, 1946.
Biologie für Mediziner. Wien, in press.
" Zur Theorie der organischen ' Gestalt '," *Roux' Arch., 108,* 1926.
" Studien über theoretische Biologie," *Biol. Zentralbl., 47,* 1927.
" Ueber die Bedeutung der Umwälzungen in der Physik für die Biologie "
(Studien über theoretische Biologie, II). *Biol. Zentralbl., 47,* 1927.
" Philosophie des Organischen " (Theoretische Biologie), *Literarische
Berichte aus dem Gebiet der Philosophie, 17/18,* 1928.
" Eduard von Hartmann und die moderne Biologie," *Arch. f. Gesch. d.
Philos. u. Soziol., 38,* 1928.
" Vorschlag zweier sehr allgemeiner biologischer Gesetze " (Studien
über theoretische Biologie, III). *Biol. Zentralbl., 49,* 1929.
" Teleologie des Lebens," *Biologia Generalis, 5,* 1929.
" Organismische Biologie," *Unsere Welt, 22,* 1930.
" Das Vitalismusproblem in ärztlicher Betrachtung," *Medizinische
Welt,* 1931.
" Tatsachen und Theorien der Formbildung als Weg zum Lebens-
problem," *Erkenntnis, 1,* 1931.
" Vaihingers Lehre von der analogischen Fiktion in ihrer Bedeutung
für die Naturphilosophie," *Vaihinger-Festschrift.* Berlin, 1932.
" Gedanken im Anschluss an neue Forschungsergebnisse über den
Bau des Protoplasmas," *Naturforscher, 10,* 1933.
" Wandlungen des biologischen Denkens," *Neue Jahrbücher,* 1934.
" Biologische Gesetzlichkeit im Lichte der organismischen Auffassung,"
Travaux du IXᵉ Congrès Internat. de Philos. (Congrès Descartes), VII.
Paris, 1937.
" Die ganzheitliche Auffassung der Lebenserscheinungen," *Kongress
für synthet. Lebensforschung.* Marienbad, 1936.
" Zu einer allgemeinen Systemlehre," *Blätter für deutsche Philosophie,
18, 3/4,* 1945.
" Zu einer allgemeinen Systemlehre," *Biologia Generalis, 19,* 1949.
" Das Weltbild der Biologie," *Europäische Rundschau,* 1948.
" Das biologische Weltbild," III. *Internationale Hochschulwochen des
österr. College in Alpbach.* Salzburg, 1948.

2. Discussions of the Organismic Conception

ALVERDES, F. : "Nochmals über die Ganzheit des Organismus," *Zool. Anz., 104,* 1933.
"Organizismus und Holismus," *Der Biologe, 5,* 1936.

BAVINK, B. : "Jenseits von Mechanismus und Vitalismus," *Unsere Welt, 21,* 1929.

BIZZARRI, A. : *Le direzioni fondamentali dei processi biologici.* Bologna, 1936.

BLEULER, E. : *Mechanismus—Vitalismus—Mnemismus.* Berlin, 1931.

BÜNNING, E. : "Mechanismus, Vitalismus und Teleologie," *Abh. d. Friesschen Schule, N. F., 5/3,* 1932.

BURKAMP, W. : "Naturphilosophie der Gegenwart," *Philos. Forschungsberichte, 2,* 1930.
Die Struktur der Ganzheiten. Berlin, 1936.
Wirklichkeit und Sinn. Berlin, 1938.

CANELLA, W. : *Orientamenti della biologia moderna.* Bologna, 1939.

CLARA, M. : *Das Problem der Ganzheit in der modernen Medizin.* Leipzig, 1940.

GESSNER, F. : "Die philosophischen Grenzfragen in der heutigen Biologie," *Freie Welt* (Gablonz), *12,* 1932.
"Theoretische Biologie," *Freie Welt* (Gablonz), *14,* 1934.

GROSS, J. : "Die Krisis in der theoretischen Physik und ihre Bedeutung für die Biologie," *Biol. Zentralbl., 50,* 1930.

HARTMANN, M. : *Philosophie der Naturwissenschaften.* Berlin, 1937.

LINSBAUER, K. : "Individuum—System—Organismus," *Mitt. Naturwiss. Verein f. Steiermark, 71,* 1934.

NEEDHAM, J. : "Thoughts on the problem of biological organization," *Scientia, 26,* 1932.

REICHENBACH, H. : *Ziele und Wege der heutigen Naturphilosophie.* Berlin, 1931.

THUMB, N. : "Die Stellung der Psychologie zur Biologie," Gedanken zu L. v. Bertalanffys Theoretischer Biologie. *Zbl. Psychotherapie, 15,* 1944.

TRIBIÑO, S. E. M. GORLERI de : *Una nueva orientación de la filosofía biológica : El organicismo de Luis Bertalanffy.* Institution Mitre de Buenos Aires. 1946.

UNGERER, E. : "Erkenntnisgrundlagen der Biologie. Ihre Geschichte und ihr gegenwärtiger Stand," *Handb. d. Biologie,* herausgegeben von L. v. Bertalanffy, Bd. I, 1941.

WENZL, A. : *Metaphysik der Biologie von heute.* Leipzig, 1938.

3. Application to Individual Problems (extract)

BERTALANFFY, L. v. : "Wesen und Geschichte des Homologiebegriffes," *Unsere Welt,* 1934.
"Untersuchungen über die Gesetzlichkeit des Wachstums. I. Allgemeine Grundlagen der Theorie ; mathematisch-physiologische Gesetzlichkeiten des Wachstums bei Wassertieren," *Roux' Arch., 131,* 1934.
Ebenso II : "A quantitative theory of organic growth," *Human Biology, 10,* 1938.

Ebenso III : " Quantitative Beziehungen zwischen Darmoberfläche und Körpergrösse bei Planaria maculata," *Roux' Arch.*, *140*, 1940.

Ebenso IV : " Probleme einer dynamischen Morphologie," *Biologia Generalis*, *15*, 1941.

Ebenso V : " Wachstumsgradienten und metabolische Gradienten bei Planarien," *Biologia Generalis*, *15*, 1941.

Ebenso VI (Mit M. RELLA): "Studien zur Reorganisation bei Süsswasserhydrozoen," *Roux' Arch.*, *141*, 1941.

Ebenso VII : " Stoffwechseltypen und Wachstumstypen," *Biol. Zbl.*, *61*, 1941.

Ebenso VIII (Mit I. MÜLLER) : " Die Abhängigkeit des Stoffwechsels von der Körpergrösse und der Zusammenhang zwischen Stoffwechseltypen und Wachstumstypen," *Rivista di Biol.*, *35*, 1943.

Ebenso IX (Mit I. MÜLLER): " Der Zusammenhang zwischen Körpergrösse und Stoffwechsel bei Dixippus morosus und seine Beziehung zum Wachstum," *Zft. vgl. Physiol.*, *30*, 1943.

Ebenso X (Mit I. MÜLLER) : " Weiteres über die Grössenabhängigkeit des Wachstums," *Biol. Zbl.*, *63*, 1943.

" Neue Ergebnisse über Stoffwechseltypen und Wachstumstypen," *Forsch. u. Fortschr. 14*, 1943.

" Metabolic types and growth types," *Research and Progress*, *9*, 1943.

" Das Wachstum in seinen physiologischen Grundlagen und seiner Bedeutung für die Entwicklung mit besonderer Berücksichtigung des Menschen," *Zft. f. Rassenkunde*, *13*, 1943.

(Mit O. HOFFMANN und O. SCHREIER) : " A quantitative study of the toxic action of quinones on Planaria gonocephala," *Nature* (London), *158*, 1946.

" Das organische Wachstum und seine Gesetzmässigkeiten," *Experientia*, *4*, 1948.

HOFFMANN-OSTENHOF, O., L. v. BERTALANFFY und O. SCHREIER : " Untersuchungen über bakteriostatische Chinone und andere Antibiotica," VII. *Monatshefte f. Chemie*, *79*, 1948.

RELLA, M. : " Vitalfärbungsuntersuchungen an Süsswasserhydrozoen," *Protoplasma*, *35*, 1940.

BERTALANFFY, L. v. : " Der Organismus als physikalisches System betrachtet," *Naturwiss.*, *28*, 1940.

" Bemerkungen zum Modell der biologischen Elementareinheiten," *Naturwiss.*, *32*, 1944.

" Vergleichende Entwicklungsgeschichte," *Handb. d. Biologie*, herausgegeben von L. v. Bertalanffy, Bd. III. In press.

4. Further Developments

BALLAUFF, Th. : " Über das Problem der autonomen Entwicklung im organischen Seinsbereich," *Blätter f. deutsche Philos.*, *14*, 1940.

" Die gegenwärtige Lage der Problematik des organischen Seins," *Blätter f. deutsche Philos.*, *17*, 1943.

BENNINGHOFF, A. : " Form und Funktion I, II," *Zft. ges Naturwiss.*, *1*, *2*, 1935/36.

" Über Einheiten und Systembildungen im Organismus," *Dtsch. med. Wschr.*, 1938.

BENNINGHOFF, A. : " Eröffnungsvortrag," *Verh. Anat. Ges.*, *46* (Erg.-H. Anat. Anz. *87*), 1939.

BURROW, P. : " The organismic factor in disorders of behaviour," *J. of Psychol.*, *4*, 1937.

DEHLINGER, U. und E. WERTZ : " Biologische Grundfragen in physikalischer Betrachtung," *Naturwiss., 30,* 1942.

DOTTERWEICH, H. : *Das biologische Gleichgewicht und seine Bedeutung für die Hauptprobleme der Biologie.* Jena, 1940.

FRIES, C. : " Wiedergeburt der Naturphilosophie," *Geistige Arbeit,* 1935. *Metaphysik als Naturwissenschaft. Betrachtungen zu L. v. Bertalanffys Theoretischer Biologie.* Berlin, 1936.

KLATT, B. : " Die theoretische Biologie und die Problematik der Schädelform," *Biologia Generalis, 19,* 1949.

LASSEN, H. : *Mechanismus, Vitalismus, Kausalgesetz a priori und die statistische Auffassung der Naturgesetzlichkeit in der modernen Physik.* Diss. Hamburg, 1931.

LEMMEL, H. : *Die Organismusidee in* MÖLLERS *Dauerwaldgedanken.* Berlin, 1939.

NATZMER, G. v. : " Individualität und Individualitätsstufen im Organismenreich," *Zft. ges. Naturwiss.,* 1935.

PRIGOGINE, I. : *Étude thermodynamique des Phénomènes irréversibles.* Liége, 1947.

PRIGOGINE, I. et J. M. WIAME : " Biologie et thermodynamique des phénomènes irréversibles," *Experientia, 2,* 1946.

REINER, J. M. and S. SPIEGELMAN : " The energetics of transient and steady states," *J. phys. Chem., 49,* 1945.

ROTHSCHUH, K. E. : *Theoretische Biologie und Medizin.* Berlin, 1936.

SPIEGELMAN, S. : " Physiological competition as a regulatory mechanism in morphogenesis," *Quart. Rev. Biol., 20,* 1945.

SYZ, H. : " The concept of the organism-as-a-whole and its application to clinical situations," *Human Biology, 8,* 1936.

WEBER, H. : " Der Umweltbegriff der Biologie und seine Anwendung," *Der Biologe, 8,* 1938.
" Zur Fassung und Gliederung eines allgemeinen biologischen Umweltbegriffes," *Naturwiss., 27,* 1939.

ZEIGER, K. : " Neuere Anschauungen über den Feinbau des Protoplasmas," *Klin. Wschr., 22,* 1943.

ZIMMERMANN, H. : " Theoretische Biologie und Heilkunde der Gegenwart," *Klin. Wschr., 11,* 1932.
" Zum Begriff des ' Biologischen ' in der Heilkunde," *Klin. Wschr., 15,* 1936.

5. *Similar Viewpoints from Other Sources*

ALVERDES, F. : " Acht Jahre tierpsychologischer Forschung im Marburger Zoologischen Institut," *Sitzber. Ges. Beförderung d. ges. Naturwiss., Marburg, 72,* 1937.

BAVINK, B. : *Ergebnisse und Probleme der Naturwissenschaften.* 8th ed., Bern, 1944.

BURTON, A. C. : " The properties of the steady state as compared to those of equilibrium as shown in characteristic biological behaviour," *J. gen. a. comp. Physiol., 14,* 1939.

CONRAD, K. : *Der Konstitutionstypus als genetisches Problem.* Berlin, 1941.

DALCQ, A. : *L'œuf et son dynamisme organisateur.* Paris, 1941.

DÜRKEN, B. : *Entwicklungsbiologie und Ganzheit.* Leipzig, 1937.

EDLBACHER, S. : " Das Ganzheitsproblem in der Biochemie," *Experientia*, *2*, 1946.

FRANKENBERG, G. : *Das Wesen des Lebens.* Braunschweig, 1933.

HIRSCH, G. Ch. : " Der Aufbau des Tierkörpers," *Handb. d. Biologie*, herausgegeben von L. v. Bertalanffy, Bd. VI, 1944.

HOLST, E. v. : " Vom Wesen der Ordnung im Zentralnervensystem," *Naturwiss.*, *25*, 1937.
" Von der Mathematik der nervösen Ordnungsfunktion," *Experientia*, *4*, 1948.

HUZELLA, Th. : *Die zwischenzellige Organisation.* Jena, 1941.

JORDAN, H. : " Die Logik der Naturwissenschaften," *Biol. Zbl.*, *52*, 1932.
Die theoretischen Grundlagen der Tierphysiologie. Leiden, 1941.

KÖHLER, O. : *Das Ganzheitsproblem in der Biologie.* Königsberg, 1930.

MITTASCH, A. : *Über katalytische Verursachung im biologischen Geschehen.* Berlin, 1935.
Über Katalyse und Katalysatoren in Chemie und Biologie. Berlin, 1936.
Katalyse und Determinismus. Berlin, 1938.

NEEDHAM, J. : *Order and Life.* Cambridge, 1936.
Integrative Levels. Oxford, 1937 (1941).

NEERGARD, K. v. : *Die Aufgabe des 20. Jahrhunderts.* 3rd ed., Erlenbach-Zürich, 1943.

OLDEKOP, E. : *Über das hierarchische Prinzip in der Natur und seine Beziehungen zum Mechanismus-Vitalismus-Problem.* Reval, 1930.

RENSCH, B. : *Neuere Probleme der Abstammungslehre.* Stuttgart, 1947.

RITTER, W. E. and E. W. BAILEY : " The organismal conception," *Univ. Calif. Publ. in Zool.*, *31*, 1928.

ROTHACKER, E. : *Die Schichten der Persönlichkeit.* 3rd ed., Leipzig, 1947.

RUSSELL, E. S. : *The Interpretation of Development and Heredity.* Oxford, 1931.

SAPPER, K. : *Biologie und organische Chemie.* Berlin, 1930.

SCHRÖDINGER, E. : *Was ist Leben?* Bern, 1946.

SEITZ, L. : *Wachstum, Geschlecht und Fortpflanzung als ganzheitlich erbmässig-hormonales Problem.* Berlin, 1939.

SKRABAL, A. : " Das Reaktionsschema der Waldenschen Umkehrung," *Österr. Chemiker-Zft.*, *48*, 1947.

UNGERER, E. : *Die Regulationen der Pflanzen.* 1st ed., Berlin, 1919.
" Erkenntnisgrundlagen der Biologie, ihre Geschichte und ihr gegenwärtiger Stand," *Handb. d. Biologie*, herausgegeben von L. v. Bertalanffy, Bd. I, 1941.

VANSELOW, K. : " Grundlagen der Forstwirtschaft," *Handb. d. Biologie*, herausgegeben von L. v. Bertalanffy, Bd. VIII/2, 1943.

WHEELER, W. M. : " Die heutigen Strömungen in der biologischen Theorie," *Unsere Welt*, *21*, 1929.

WOLTERECK, R. : *Ontologie des Lebendigen.* Stuttgart, 1940.

WOODGER, J. H. : *Biological Principles.* London, 1929.

6. *Recent Work Pertinent to Organismic Biology*

BALLAUFF, TH. : *Das Problem des Lebendigen.* Bonn, 1949.

BENTLEY, A. F. : " Kennetic inquiry," *Science 112*, 1950.

BERTALANFFY, L. von : " Problems of organic growth," *Nature, 163,* 1949.
" Goethes Naturauffassung," *Atlantis* (Zurich), *8,* 1949. " Goethe's concept of nature," *Main Currents in Modern Thought, 8,* 1951.
" The theory of open systems in physics and biology," *Science, 111,* 1950.
" An outline of General System Theory," *Brit. J. Philos. Sci., 1,* 1950.
" Growth types and metabolic types," *Amer. Naturalist, 85,* 1951.
" Theoretical models in biology and psychology," In : Theoretical Models and Personality Theory. (Symposium.) *J. Personality,* 1951.

BERTALANFFY, L. von, C. G. HEMPEL, R. E. BASS, and H. JONAS : " General System Theory—A new approach to unity of science " (Symposium), *Human Biology, 1951.*

BERTALANFFY, L. von, and W. J. PIROZYNSKI : " Tissue respiration and body size," *Science, 113,* 1951.

BODE, H., *et al.* : " The education of a scientific generalist," *Science, 109,* 1949.

BRUNSWIK, E. : " The Conceptual Framework of Psychology," *Internat. Encyclopedia of Unified Science,* vol. 1, No. 10. Chicago, 1950. (Preliminary mimeographed edition.)

CANTRIL, H., *et al.* : " Psychology and scientific research," *Science, 110,* 1949.

HOBBIGER, F. und G. WERNER : " Ueber das chemische Gleichgewicht der Acetylcholinverteilung im Gehirngewebe von Warmbluetern," *Z. Vitamin-Hormon-Fermentforsch, 2,* 1949.
" Zum Verhalten der Acetylcholinsynthese im Gehirnbrei von Warmbluetern," *Arch. int. Pharmacodyn., 79,* 1949.

KRECH, D. : " Dynamic systems as open neurological systems," *Psychol. Rev., 57,* 1950.

LUDWIG, W. und J. KRYWIENCZYK : " Koerpergroesse, Koerperzeiten und Energiebilanz," III. *Z. vgl. Physiol., 32,* 1950.

MENNINGER, K. : *Psychiatric Nosology and the Nature of Illness.* (Mimeograph.) Topeca (Ka.), 1951.

NETTER, H. : " Die Feinstruktur der Zelle als dynamisches Geschehen," *Verh. Dtsch. Path. Ges., 1949.*

SCHREIER, O. : " Die schaedigende Wirkung verschiedener Chinone auf Planaria gonocephala Dug. und ihre Beziehung zur Childschen Gradiententheorie," *Oesterr. Zool. Z., 2,* 1949.

SCHULZ, G. V. : " Ueber den makromolekularen Stoffwechsel der Organismen," *Naturwiss., 37,* 1950.

SEIDEL, F. : *Goethe gegen Kant.* Berlin, 1948.

SKRABAL, A. : " Die Kettenreaktionen, anders gesehen," *Mh. Chemie, 80,* 1949.

TRIBIÑO, S. MORALES GORLERI de : " Las conquistas de la Fisica y su repercusión en la Biologia. Las teorias de Schrödinger y de Bertalanffy sobra la estructura de los chromosomas," *Anales Soc. Cientif. Argentina, entrega I, 146,* 1948.

WERNER, G. : " Beitrag zur mathematischen Behandlung pharmakologischer Fragen," *Sitzber. Akad. Wiss. Wien, Math.-naturwiss. Kl. Abt. IIa. 156,* 1947.

ZIMMERMANN, H. : " Das Gefuege der Heilkunde," *Med. Klinik, 1946.*

7. Suggested Readings on Modern Biological Thought

ALEXANDER, J.: *Life. Its Nature and Origin.* New York, 1948.

CANNON, W. B.: *The Wisdom of the Body.* New York, 1939.

CARREL, A.: *Man The Unknown.* New York and London, 1939.

DRIESCH, H.: *The Science and Philosophy of the Organism.* New York, 1928.

HOLMES, S. J.: *Organic Form and Related Biological Problems.* Berkeley, 1940.

HUXLEY, J.: *Evolution. The Modern Synthesis.* 4th impr. London, 1945.

LECOMTE DU NOÜY: *Human Destiny.* New York, 1947.

LILLIE, R. S.: *General Biology and Philosophy of the Organism.* Chicago, 1945.

NEEDHAM, J.: *Order and Life.* New Haven, 1936.

RUSSELL, E. S.: *The Directiveness of Organic Activities.* Cambridge (England), 1945.

SCHRÖDINGER, E.: *What is Life?* New York, 1946.

SHERRINGTON, Sir CHARLES S.: *Man and His Nature.* Cambridge (England), 1945.

SIMPSON, G. G.: *Tempo and Mode in Evolution.* New York, 1944.

SINGER, CH.: *A History of Biology.* London, 1950.

SINNOTT, E. M.: *Cell and Psyche. The Biology of Purpose.* Chapel Hill, 1950.

WIENER, N.: *Cybernetics.* New York and Paris, 1948.

WOODGER, J. H.: *Biological Principles.* New York and London, 1929.

INDEX OF AUTHORS

(Italic figures indicate authors in References)

INDEX OF SUBJECTS